STATISTICS for CRIMINOLOGY and CRIMINAL JUSTICE

To Aaron, Razberri, Norm, Hank, Lola, Pip, and Bus for all your love and support. Life is good because of you.

STATISTICS for CRIMINOLOGY and CRIMINAL JUSTICE

JACINTA M. GAU

University of Central Florida

Los Angeles | London | New Delhi
Singapore | Washington DC

Los Angeles | London | New Delhi
Singapore | Washington DC

FOR INFORMATION:

SAGE Publications, Inc.
2455 Teller Road
Thousand Oaks, California 91320
E-mail: order@sagepub.com

SAGE Publications Ltd.
1 Oliver's Yard
55 City Road
London EC1Y 1SP
United Kingdom

SAGE Publications India Pvt. Ltd.
B 1/I 1 Mohan Cooperative Industrial Area
Mathura Road, New Delhi 110 044
India

SAGE Publications Asia-Pacific Pte. Ltd.
33 Pekin Street #02-01
Far East Square
Singapore 048763

Executive Editor: Jerry Westby
Editorial Assistant: Laura Cheung
Production Editor: Brittany Bauhaus
Copy Editor: Alan Cook
Typesetter: C&M Digitals (P) Ltd.
Proofreader: Theresa Kay
Indexer: Kathy Paparchontis
Cover Designer: Karine Hovsepian
Marketing Manager: Jordan Bell
Permissions Editor: Karen Ehrmann

Printed in the United States of America

Library of Congress Cataloging-in-Publication Data

Gau, Jacinta M., 1982–

Statistics for criminology and criminal justice / Jacinta M. Gau.

p. cm.
Includes bibliographical references.

ISBN 978-1-4129-9127-8 (pbk.)

1. Criminal statistics. 2. Statistical methods. I. Title.

HV7415.G38 2013
519.5—dc23   2011050630

This book is printed on acid-free paper.

SFI® Certified Sourcing
www.sfiprogram.org
SFI-00453

12 13 14 15 16 10 9 8 7 6 5 4 3 2 1

Brief Contents

DETAILED CONTENTS

PREFACE

As a statistics instructor, I hear it from my students time and again: "I'm not a math person" and "I just don't *get* numbers." I consider it my foremost obligation to disabuse them of such misconceptions. These mental blocks ultimately produce self-fulfilling prophecies when students perform poorly in the course and leave the room after the final exam having learned nothing about statistics but feeling reaffirmed in the accuracy of their self-described innumeracy. Statistics should be an enlightening and eye-opening course, not one that students feel persistently highlights their intellectual weak points. Many of my students, coming to pick up their graded final exams after their term in the class has ended, have remarked to me, "This class wasn't nearly as bad as I thought it would be." A quasicompliment to professors of other subjects, perhaps, this is an important statement coming from statistics students, because it demonstrates that their preconceived notions about their academic inadequacies can be falsified. Key to successful falsification is the use of a good textbook.

A good statistics book is one that balances the quantity and the complexity of information presented in its pages. Skimping on the details for the sake of simplicity can create a superficial text that is too sparse. Many of the details sacrificed may pertain to the underlying theory and logic that drive statistical analyses. While some might argue that the logical underpinnings of statistics are too complicated for novice undergraduates, my thesis is that students have to know the logic in order to understand the math. Without the theory of probability, the logic of hypothesis testing, and the understanding of their practical application, statistics is mere form of numbers and students will resort to trying to memorize rather than trying to truly understand.

Of course, an overly complex textbook can have an eye-glazing effect. An overdose of details is tantamount to throwing students into the deep end: Many sink immediately, a few float without trouble, and the remainder hover at near-drowning levels for the 10 or 15 weeks it takes for the term to end. The amount of information presented in such a textbook, moreover, may be unrealistic given the time constraints of semesters and, in particular, quarters.

Statistics for Criminology and Criminal Justice attempts to strike a balance among depth, breadth, and accessibility. The pedagogical aim of this text is to limit jargon and present only that information necessary to the comprehension of the material being presented while still including coverage of the theoretical fundamentals such as probability and sampling distributions. The goal is to pare the information down enough such that students do not get lost in the details yet retain information vital to the understanding of theory and logic.

That said, instructors who wish to do so can skip the sampling distribution chapter and portions of the probability chapter. The last chapter of the book, which introduces bivariate and multiple regression, can also be dropped pursuant to instructor discretion and time constraints. This book is designed to be flexible so that a wide variety of instructors and students can be accommodated.

This book is set apart from similar texts designed for criminal justice and criminology by its use of article summaries and real data. Where relevant, *Research Example* boxes appear throughout the chapters. These examples are short summaries of peer-reviewed journal articles that used one of the techniques covered in the chapter, so that students can see the practical application of that technique. The topic coverage is broad so that students can experience the types of questions that researchers ask and can gain a sense for how interesting and diverse these questions are.

Chapters also contain *Data Sources* boxes that describe some common, publicly available data sets such as the Uniform Crime Reports, National Crime Victimization Survey, State Court Processing Statistics, General Social Survey, and others. All in-text examples and most end-of-chapter review problems utilize real data drawn from the data sources highlighted in the book. The goal is to lend a practical, tangible bent to the often-abstract topic of statistics. Students get to work with the data that their professors use. They get to see how elegant statistics can be at times and how messy they can be at others, how analyses can sometimes lead to clear conclusions and other times end in ambiguity. These are lessons that no list of numbers fabricated purely for the sake of example can provide.

Each chapter ends with a section on SPSS and comes with one or more pared-down versions of a major data set in SPSS format. Students can download these small data sets to answer the review questions presented at the end of the chapter, and any other study questions instructors may wish to assign. The full data sets are all available from the Inter-University Consortium for Political and Social Research at www.icpsr.umich.edu/icpsrweb/ICPSR/. If desired, instructors can download the data sets and provide supplementary examples and practice problems for hand calculations or SPSS analyses.

The book is presented in three parts. Part I covers descriptive statistics. It starts with the very basics of levels of measurement and moves on to frequency distributions, graphs and charts, proportions and percentages, central tendency, and dispersion. Part II focuses on probability theory and sampling distributions. As mentioned above, instructors can assign this entire section or can cut out chapters or portions of chapters as they prefer. Part II ends with confidence intervals, which provide students' first foray into inferential statistics.

Part III begins with an introduction to bivariate hypothesis testing, which is intended to ease students into inferential tests by explaining just what it is that these tests do and what they are for. The book then covers chi-square tests, *t* tests, ANOVA, correlation, and introductory ordinary least squares regression. The order of these chapters is not accidental; the sequence is designed such that some topics flow logically into others. Chi-square tests are presented first because they are the only nonparametric test type covered here and, more importantly, they are probably the least computationally intense of the analyses and therefore provide a relatively nonthreatening initial experience with hypothesis testing. Two-population *t* tests are the subject of the following chapter, and they flow logically into ANOVA in the proceeding chapter. Correlation, likewise, segues into regression, though the latter is not required and the correlation chapter can stand alone.

I hope that this book offers an attractive compromise between depth and straightforwardness and that the use of articles, data sets, and SPSS provides a hands-on experience that makes statistics more tangible as well as more vivid, interesting, and alive. The presentation style and the step-by-step display of calculations is intended to help convince students to let go of their "math block" presumptions in favor of the belief that they can succeed in the course. It is my wish that this book may facilitate learning in a manner such that students leave the course with a solid comprehension of the basics of statistics, a large hole punched into their beliefs about their inability to "do" math, and the belief that although they feared the class, it turned out not to be nearly as bad as they thought it would be.

ACKNOWLEDGMENTS

I am indebted to Leana Bouffard, currently Associate Professor in the College of Criminal Justice at Sam Houston State University, for not only educating me about statistics but for teaching me how to teach them. It was through observing her method of conveying information to students—undergraduate and graduate alike—that I learned the keys to effectively communicating this complex topic. I also owe heartfelt thanks to Travis Pratt, currently Professor in the School of Criminology and Criminal Justice at Arizona State University, for always being ready to share his quantitative expertise with others and for his patience with my never-ending streams of questions; and to Craig Parks and G. Leonard Burns, both Professors in the Department of Psychology at Washington State University, who translate their extensive knowledge of statistics into informative, dynamic university courses. I wish to express my gratitude to Mr. Harry O'Dell, my high school math teacher, who took his subject seriously, demanded high performance, and told me I could attend any university I wanted. I still have my TI-86. Finally, the list of acknowledgments would be incomplete without recognition of the unfailing support that my mom, Eileen, has given all of her children. She is without doubt the most committed and patient person I have ever met.

With regard to the development and preparation of this manuscript, I wish to thank Jerry Westby and the staff at SAGE for their support and encouragement. You guys are the best! Numerous reviewers also provided advice, recommendations, and, yes, even criticisms that helped shape this book. They are listed in alphabetical order below. Of course, in spite of all the input from numerous individuals, any errors contained in this text are mine alone.

Lori Anderson, *Tarleton State University*
Vivian Andreescu, *University of Louisville*
Brenda Sims Blackwell, *Georgia State University*
Jeb Booth, *Salem State College*
Michael J. DeValve, *Fayetteville State University*
Steve Ellwanger, *East Tennessee State University*
Shannon Fowler, *University of Texas - Arlington*
Lisa Graziano, *California State University - Los Angeles*
Suman Kakar, *Florida International University*
Zachary Hamilton, *Washington State University*
Wendy L. Hicks, *Loyola University - New Orleans*
Jim Holstein, *Marquette University*
Dae Hoon, *Texas A & M International University*
Anthony Hoskin, *University of Texas - Permian Basin*

Beth M. Huebner, *University of Missouri*
David Hull, *Saint Anselm College*
Connie Ireland, *California State University - Long Beach*
Wesley G. Jennings, *University of South Florida*
Shanhe Jiang, *University of Toledo*
Rebecca S. Katz, *Morehead State University*
Carlos E. Posadas, *New Mexico State University*
Benjamin Steiner, *University of South Carolina*
Brian Stults, *Florida State University*
Christopher Sullivan, *University of Cincinnati*
Angela P. Taylor, *Fayetteville State University*
Carleen Vincent, *Florida International University*
Edward Wells, *Illinois State University*

About the Author

Jacinta M. Gau received her PhD from Washington State University in 2008. She is currently Assistant Professor in the Department of Criminal Justice at the University of Central Florida. Her work has appeared in journals such as *Justice Quarterly*, *Crime & Delinquency*, *Criminology & Public Policy*, *Police Quarterly*, *Policing: An International Journal of Police Strategies and Management*, the *Journal of Criminal Justice*, and the *Journal of Criminal Justice Education*. She is also coauthor, with Drs. Travis Pratt and Travis Franklin, of the book *Key Ideas in Criminology and Criminal Justice,* published by SAGE.

PART I

DESCRIPTIVE STATISTICS

Chapter 1

Introduction to the Use of Statistics in Criminal Justice and Criminology

Y ou might be thinking, "What do statistics have to do with criminal justice or criminology?" It is perfectly reasonable for you to question the requirement that you spend an entire term poring over a book about statistics instead of one about policing, courts, corrections, or criminological theory. Many criminology/criminal justice undergraduates wonder, "Why am I here?" In this context, the question is not so much existential as it is practical; luckily, the answer is equally practical.

You are "here" (in a statistics course) because the answer to the question of what statistics have to do with criminal justice/criminology is "Everything!" Statistical methods are the backbone of criminal justice/criminology as a field of scientific inquiry. Statistics enable the construction and expansion of knowledge about criminality and the criminal justice system. Research that tests theories or examines criminal justice phenomena and is then published in the form of academic journal articles and books is the basis for most of what we know about criminal offending and the system that has been designed to deal with it. The majority of this research would not be possible without statistics.

RESEARCH EXAMPLE 1.1

What Do Criminal Justice/Criminology Researchers Study?

Researchers in the field of criminology and criminal justice examine a wide variety of issues pertaining to the criminal justice system and theories of offending. Included are things such as prosecutorial charging decisions, racial and gender disparities in

---

sentencing, police use of force, drug and domestic violence courts, and recidivism. The following are examples of studies that have been conducted and published. You can find the full text of each of these articles and of all those presented in the following chapters at (**http://www.sagepub.com/gau**).

1. *Can an anticrime strategy that has been effective at reducing certain types of violence also be used to combat open-air drug markets?* The "pulling levers" approach involves deterring repeat offenders from crime by targeting them for enhanced prosecution while also encouraging them to change their behavior by offering them access to social services. This strategy has been shown to hold promise with gang members and others at risk for committing violence, so the Rockford (Illinois) Police Department (RPD) decided to find out if they could use a pulling levers approach to tackle open-air drug markets and the crime problems caused by these nuisance areas. After the RPD implemented the pulling levers intervention, Corsaro, Brunson, and McGarrell (2009) used official crime data from before and after the intervention to determine whether this approach had been effective. They found that although there was no reduction in violent crime, nonviolent crime (for example, drug offenses, vandalism, and disorderly conduct) declined noticeably after the intervention. This indicated that the RPD's efforts had worked, because drug and disorder offenses were exactly what the police were trying to reduce.

2. *Are prisoners with low self-control at heightened risk of victimizing, or being victimized by, other inmates?* Research has consistently shown that low self-control is related to criminal offending. Some studies have also indicated that this trait is a risk factor for victimization, in that people with low self-control may place themselves in dangerous situations. One of the central tenets of this theory is that self-control is stable and acts in a uniform manner regardless of context. Kerley, Hochstetler, and Copes (2009) tested this theory by examining whether the link between self-control and both offending and victimization held true within the prison environment. Using data gathered from surveys of prison inmates, the researchers discovered that low self-control was only slightly related to in-prison offending and victimization. This result may call into question the assumption that low self-control operates uniformly in all contexts; to the contrary, something about prisoners themselves, the prison environment, or the interaction between the two may change the dynamics of low self-control.

3. *How prevalent is victim precipitation in intimate partner violence?* A substantial portion of violent crimes are initiated by the person who ultimately becomes the victim in an incident. Muftić, Bouffard, and Bouffard (2007) explored the role of victim precipitation in instances of intimate partner violence (IPV). They gleaned data from IPV arrest reports and found that victim precipitation was present in

(Continued)

(Continued)

cases both of male and female arrestees, but that it was slightly more common in instances where the woman was the one arrested. This suggested that some women (and, indeed, some men) arrested for IPV might be responding to violence initiated by their partners rather than themselves being the original aggressors. The researchers also discovered that victim precipitation was a large driving force behind dual arrests (cases in which both parties are arrested), as police could either see clearly that both parties were at fault or, alternatively, were unable to determine which party was the primary aggressor. Victim precipitation and the use of dual arrests, then, could be contributing factors behind the recent rise in the number of women arrested for IPV against male partners.

4. *Does having a close personal friend who is black influence whites' perceptions of crime as a problem?* The racialization of violent crime has been argued to have indelibly linked blacks with predatory violence and spawned a general antipathy toward blacks among white Americans. Mears, Mancini, and Stewart (2009) hypothesized that whites who had at least one close friend who was black would be less concerned about crime because their personal exposure to blacks would cause a breakdown in the stereotype about blacks as crime-prone. The researchers examined survey data from white respondents and actually found the opposite of what they had predicted—rather than lessening whites' concern about crime, having a black friend *heightened* that concern. The researchers suggested that having a black friend may increase whites' exposure to and knowledge about the dangerous urban environment that many blacks must navigate daily and thereby elevate whites' concern about local and national crime.

5. *Against whom do police use Tasers? Why do they use them? Are Tasers effective?* Tasers and other conducted energy devices have proliferated over the past few years, yet research has been slow to address this important and controversial addition to the police toolkit. White and Ready (2007) sought to help fill this void by examining the characteristics of suspects, officers, and situations in which Tasers were deployed. They used data on Taser-involved incidents from a large metropolitan police agency. They found that the most frequently cited reason for Taser use by officers was that a suspect was emotionally disturbed and was exhibiting violent behavior, either self-directed or directed toward the police. Over one half of suspects were unarmed, and nearly all were transported to a hospital afterward for medical examinations. Most suspects were immediately incapacitated by the jolt and ceased resistance after one application, leading to overwhelmingly positive reports by officers regarding the Taser's effectiveness at defusing dangerous situations quickly and safely.

Statistics can be very abstract, so this book uses two techniques to add a realistic, pragmatic dimension to the subject. The first technique is the use of examples of statistics in criminal justice/criminology research. These summaries are contained in the *Research Example* boxes embedded in each chapter. They are meant to give you a glimpse into the types of questions that are asked in this field of research and the ways in which specific statistical techniques are utilized to answer those questions. You will see firsthand how lively and diverse criminal justice/criminology is. Research Example 1.1 summarizes five studies. Take a moment now to read through them.

The second method is the use of real data from reputable and widely used sources such as the Bureau of Justice Statistics (BJS). BJS is housed within the U.S. Department of Justice and is responsible for gathering, maintaining, and analyzing data on various criminal justice topics at the county, state, and national levels. Visit http://bjs.ojp.usdoj.gov/ to familiarize yourself with BJS. The purpose behind the use of real data is to give you the type of hands-on experience that you cannot get from fictional numbers. You will come away from this book having worked with some of the same data that criminal justice/criminology researchers use. Two sources of data that will be used in upcoming chapters are the Uniform Crime Reports (UCR) and the National Crime Victimization Survey (NCVS). See Data Sources 1.1 and 1.2 for information about these commonly used measures of offending and victimization, respectively. All of the data sets used in this book are publicly available and were downloaded from the archive maintained by the Inter-University Consortium for Political and Social Research at www.icpsr.umich.edu.

DATA SOURCES 1.1

THE UNIFORM CRIME REPORTS (UCR)

The Federal Bureau of Investigation (FBI) collects annual data on crimes reported to police agencies nationwide and maintains the Uniform Crime Reports (UCR). Crimes are sorted into eight index offenses: homicide, rape, robbery, aggravated assault, burglary, larceny-theft, motor vehicle theft, and arson. An important aspect of this data set is that it only includes those crimes that come to the attention of police—crimes that are not reported or otherwise detected by police are not counted. The UCR also conforms to the hierarchy rule, which mandates that in multiple-crime incidents, only the most serious offense ends up in the UCR. If, for example, someone breaks into a residence with intent to commit a crime inside the dwelling and while there, he kills the homeowner and then sets fire to the structure to hide the crime, he has committed burglary, murder, and arson. Because of the hierarchy rule, though, only the murder would be reported to the FBI—it would be as if the burglary and arson had never occurred. Due to underreporting by victims and the hierarchy rule, then, the UCR undercounts the amount of crime in the United States. It nonetheless offers valuable information and is widely used. You can explore this data source at www.fbi.gov/about-us/cjis/ucr/ucr.

DATA SOURCES 1.2

THE NATIONAL CRIME VICTIMIZATION SURVEY (NCVS)

The Bureau of the Census conducts the periodic NCVS under the auspices of BJS to estimate the number of criminal incidents that transpire per year and to collect information about crime victims. Multistage cluster sampling is used to select a random sample of households, and each member of that household who is 12 or older is asked to participate in an interview. Those who agree to be interviewed are asked over the phone or in person about any and all criminal victimizations that transpired in the 6 months prior to the interview. The survey employs a rotating panel design, so respondents are called at 6-month intervals for a total of 3 years, and then new respondents are selected (Bureau of Justice Statistics, 2006). The benefit of the NCVS over the UCR is that NCVS respondents might disclose victimizations to interviewers that they did not report to police, thus making the NCVS a better estimation of the total volume of crime in the U.S. The NCVS, though, suffers from the weakness of being based entirely upon victims' memory and honesty about the timing and circumstances surrounding criminal incidents. The NCVS also excludes children younger than 12, institutionalized populations (for example, persons in prisons, nursing homes, and hospitals), and the homeless. Despite these problems, the NCVS is useful because it facilitates research into the characteristics of crime victims.

In this book, emphasis will be placed on both the production and consumption of statistics. Every statistical analysis has a "producer" (someone who runs the analysis) and a "consumer" (someone to whom an analysis is being presented). Regardless of which role you play in any given situation, it is vital that you are sufficiently versed in quantitative methods that you can identify the proper statistical technique and correctly interpret the results. When you are in the consumer role, you must also be ready to question the methods used by the producer so that you can determine for yourself how trustworthy the results are. Critical thinking skills are an enormous component of statistics. You are not a blank slate standing idly by, waiting to be written upon—you are an active agent in your acquisition of knowledge about criminal justice, criminology, and the world in general. Be critical, be skeptical, and never hesitate to ask for more information.

▣ SCIENCE: BASIC TERMS AND CONCEPTS

There are a few terms and concepts that you must know before you get into the substance of the book. Statistics are a tool in the larger enterprise of scientific inquiry. **Science** is the process of gathering information and developing knowledge using techniques and procedures that are accepted by other scientists in a discipline. Science is grounded in **methods**—research results are trustworthy only when the procedures used to reach them are considered correct by others in the scientific community. To this end, everybody who conducts a study bears an obligation to be very clear and open about the methods they used. It is never proper to question scientific results on the basis of a moral, emotional, or opinionated disagreement with them, but it is entirely correct to question results when the procedures used to arrive at them are shoddy or inadequate.

Science: The process of gathering and analyzing data in a systematic and controlled way using procedures that are generally accepted by others in the discipline.

Methods: The procedures used to gather and analyze scientific data.

STUDY TIP: GIGO!

Part of the methodological consideration involves the process employed to gather data—this is the topic covered in research methods courses. The other part involves the techniques used to analyze the data to look for patterns and test for relationships—this is the realm of statistics. Together, proper methods of gathering and analyzing data form the groundwork for scientific inquiry. If there is a flaw in either the gathering or the analyzing of data, then the results may not be trustworthy. "Garbage in, garbage out" (GIGO) is the mantra of statistics. Data gathered with the best of methods can be rendered worthless if the wrong statistical analysis is applied to them; likewise, the most sophisticated, cutting-edge statistical technique cannot salvage improperly collected data. When the data or the statistics are defective, the results are likewise deficient and cannot be trusted.

A key aspect of science is the importance of **replication**. No single study ever "proves" something definitively; quite to the contrary, much testing must be done before firm conclusions can be drawn. Replication is important because there are times when a study is flawed and needs to be redone, or when the original study is methodologically sound but an alteration of the statistical procedure produces different results. The scientific method's requirement that all researchers divulge the steps they took to gather and analyze data allows other researchers and members of the public to examine those steps and, if warranted, to undertake replications.

Replication: The repetition of a particular study that is conducted for purposes of determining whether the original study's results hold when new samples or measures are employed.

▣ TYPES OF SCIENTIFIC RESEARCH IN CRIMINAL JUSTICE AND CRIMINOLOGY

Criminal justice/criminology research is diverse in nature and purpose. Much of it involves theory testing. **Theories** are proposed explanations for certain events. **Hypotheses** are small "pieces" of theories that must be true in order for the entire theory to hold up. You can think of a theory as a chain and hypotheses as the links forming that chain. In Research Example 1.1, Kerley et al. (2009) conducted a test of the general theory of crime. The general theory holds that low self-control is a static predictor of offending and victimization, regardless of context. From this proposition, the researchers deduced the hypothesis that the relationship between low self-control and both offending and victimization must hold true in the prison environment. Their results showed an overall lack of support for the hypothesis that low self-control operates uniformly in all contexts, thus calling that aspect of the general theory of crime into question. This is an example of a study designed to test a theory.

Theory: A set of proposed and testable explanations about reality that are bound together by logic and evidence.

Hypothesis: A single proposition, deduced from a theory, that must hold true in order for the theory itself to be considered valid.

Evaluation research is also common in criminal justice/criminology. In Research Example 1.1, the article by Corsaro et al. (2009) is an example of evaluation research. This type of study is undertaken when a new policy, program, or intervention is put into place and researchers want to know whether the intervention accomplished its intended purpose. In this study, the Rockford Police Department implemented a pulling levers approach to combat drug and nuisance offending. After the program had been put into place, the researchers analyzed crime data to find out whether or not the approach was effective.

Exploratory research occurs when there is limited knowledge about a certain phenomenon; researchers essentially embark into unfamiliar territory when they attempt to study this social event. The study by Muftić et al. (2007) in Research Example 1.1 was exploratory in nature because so little is known about victim precipitation, particularly in the realm of intimate partner violence. It is often dangerous to venture into new areas of study when the theoretical guidance is spotty; however, exploratory studies have the potential to open new areas of research that have been neglected but that provide rich information that expands the overall body of knowledge.

Finally, some research is **descriptive** in nature. White and Ready's (2007) analysis of Taser deployments exemplifies a descriptive study. They did not set out to test a theory or explore a new area of research—they merely offered basic descriptive information about the suspects, officers, and situations involved in Taser incidents. This type of research can be very informative when knowledge about a particular phenomenon is scant.

Evaluation research: Studies intended to assess the results of programs or interventions for purposes of discovering whether those programs or interventions appear to be effective.

Exploratory research: Studies that address issues that have not been examined much or at all in prior research and that therefore may lack firm theoretical and empirical grounding.

Descriptive research: Studies done solely for the purpose of describing a particular phenomenon as it occurs in a sample.

With the exception of purely descriptive research, the ultimate goal in most statistical analyses is to generalize from a **sample** to a **population**. A population is the entire set of people, places, or objects that a researcher wishes to study. Populations, though, are usually very large. Consider, for instance, a researcher trying to estimate attitudes about capital punishment in the general U.S. population. That is a population of more than 300 million! It would be impossible to measure everyone directly. Researchers thus draw samples from populations and study the samples instead. **Probability sampling** helps ensure that a sample mirrors the population from which it was drawn (for example, a sample of people should contain a breakdown of race, gender, and age similar to that found in the population). Samples are smaller than populations, and researchers are therefore able to measure and analyze them. The results found in the sample are then generalized to the population.

Sample: A subset pulled from a population with the goal of ultimately using the people, objects, or places in the sample as a way to generalize to the population.

Population: The universe of people, objects, or locations that researchers wish to study. These groups are often very large.

Probability sampling: A sampling technique in which all people, objects, or areas in a population have an equal and known chance of being selected into the sample.

▣ SOFTWARE PACKAGES FOR STATISTICAL ANALYSIS

Hand computations are the foundation of this book because it is through seeing the numbers and working with the formulae that you gain an understanding of statistical analyses. In the real world, however, statistical analysis is generally conducted using a software program. Microsoft Excel contains some rudimentary statistical functions and is commonly used in situations where the only things sought are basic descriptive analyses; however, this program's usefulness is exhausted quickly because researchers usually want far more than descriptives. There are many statistical packages available. The most common in criminal justice/criminology research are SPSS (Statistical Package for the Social Sciences), Stata, and SAS (Statistical Analysis Software).

This book incorporates SPSS into each chapter. This allows you to get a sense for what data look like when displayed in their raw format, for how to run particular analyses, and for how to read and interpret program output. Where relevant, the chapters offer SPSS practice problems and accompanying data sets that are available for download from **(http://www.sagepub.com/gau)**. This offers a practical, hands-on lesson about the way that criminal justice/criminology researchers use statistics.

▣ ORGANIZATION OF THE BOOK

This book is divided into three parts. Part I covers descriptive statistics. Chapter 2 provides a basic overview of types of variables and levels of measurement. Some of this material will be review for students who have taken a methods course. Chapter 3 delves into charts and graphs as means of graphically displaying data. Measures of central tendency are the topic of Chapter 4. These are descriptive statistics that let you get a feel for where the data are clustered. Chapter 5 discusses measures of dispersion. Measures of dispersion complement measures of central tendency by offering information about whether the data tend to cluster tightly around the center or whether they are very spread out.

Part II describes the theoretical basis for statistics in criminal justice and criminology: probability and probability distributions. Part I of the book can be thought of as the nuts-and-bolts of the mathematical concepts used in statistics, and Part II can be seen as the theory behind the math. Chapter 6 introduces probability theory. Binomial and continuous probability distributions are discussed. In Chapter 7, you will learn about population, sample, and sampling distributions. Chapter 8 provides the book's first introduction to inferential statistics with its coverage of point estimates and confidence intervals. The introduction of inferential statistics at this juncture is designed to help ease you into Part III.

Part III of the book merges the concepts learned in Parts I and II to form the discussion on inferential hypothesis testing. Chapter 9 offers a conceptual introduction to this framework, including a description of the five steps of hypothesis testing that will be used in every proceeding chapter. In Chapter 10, you will encounter your first bivariate statistical technique: chi-square. Chapter 11 describes two-population t tests and tests for differences between proportions. Chapter 12 covers analysis of variance, which is an extension of the two-population t test. In Chapter 13, you will learn about correlations. Finally, Chapter 14 wraps up the book with an introduction to bivariate and multiple regression.

The prerequisite that is indispensible to success in this course is a solid background in algebra. You absolutely must be comfortable with basic techniques such as adding, subtracting, multiplying, and dividing. You also need to understand the difference between positive and negative numbers. You will be required to plug numbers into equations and solve those equations. These are procedures that you should not have a problem with if you remember the lessons you learned in your high school and college algebra courses. Appendix A offers an overview of the algebraic concepts you will need, so look those over and make sure that you are ready to take this course.

Statistics are cumulative in that many of the concepts you learn at the beginning form the building blocks for more complex techniques that you will learn about as the course progresses. Means, proportions, and standard deviations, for instance, are things you will learn about in Part I, but they will remain relevant throughout the remainder of the book. You must, therefore, learn these fundamental calculations well and you must remember them.

Repetition is the key to learning statistics. Practice, practice, practice! There is no substitute for doing and redoing the end-of-chapter review problems and any other problems your instructor may provide. You can also use the in-text examples as problems if you just copy down the numbers and do the calculations on your own without looking at the book. Remember, even the most advanced statisticians started off knowing nothing about statistics. The learning process is something everyone has to go through. You will complete this process successfully as long as you have basic algebra skills and are willing to put in the time and effort it takes to succeed.

CHAPTER 1 REVIEW PROBLEMS

1. Define science and explain the role of methods in the production of scientific knowledge.

2. Name three theories that you have encountered in your criminal justice/criminology classes. For each one, write one hypothesis that you could gather data on and test.

3. What is a population? Why are researchers usually unable to study populations directly?

4. What is a sample? Why do researchers draw samples?

5. Explain the role of replication in science.

6. What does *GIGO* stand for? What does this acronym mean in the context of statistical analyses?

KEY TERMS

Science	Hypothesis	Sample
Methods	Evaluation research	Population
Replication	Exploratory research	Probability sampling
Theory	Descriptive research	

REFERENCES

Bureau of Justice Statistics. (2006). *National crime victimization survey, 2004[record-type files]: Codebook.* Washington, DC: U.S. Department of Justice.

Corsaro, N., Brunson, R. K., & McGarrell, E. F. (2009). Problem-oriented policing and open-air drug markets: Examining the Rockford pulling levers deterrence strategy. *Crime & Delinquency.* Prepublished October 14, 2009. DOI: 10.1177/0011128709345955

Kerley, K. R., Hochstetler, A., & Copes, H. (2009). Self-control, prison victimization, and prison infractions. *Criminal Justice Review, 34*(4), 553–568.

Mears, D. P., Mancini, C., & Stewart, E. A. (2009). Whites' concern about crime: The effects of interracial contact. *Journal of Research in Crime and Delinquency, 46*(4), 524–552.

Muftić, L. R., Bouffard, L. A., & Bouffard, J. A. (2007). An exploratory analysis of victim precipitation among men and women arrested for intimate partner violence. *Feminist Criminology, 2*(4), 327–346.

White, M. D., & Ready, J. (2007). The TASER as a less lethal force alternative: Findings on the use and effectiveness in a large metropolitan police agency. *Police Quarterly, 10*(2), 170–191.

Chapter 2

Types of Variables and Levels of Measurement

The first thing you must be familiar with in statistics is the concept of a **variable**. A variable is, quite simply, something that varies. It is a coding scheme used to measure a particular characteristic of interest. If, for instance, you asked all of your statistics classmates, "How many classes are you taking this term?" you would receive many different answers. This would be a variable. Variables sit in contrast to **constants**, which are characteristics that assume only one value in a sample. It would be pointless for you to ask all of your fellow classmates whether they are taking statistics this term because of course the answer they would all provide is "yes."

Variable: A characteristic that describes people, objects, or places and takes on multiple values in a sample or population.

Constant: A characteristic that describes people, objects, or places and takes on only one value in a sample or population.

RESEARCH EXAMPLE 2.1

Choosing Variables for a Study on Police Use of Conductive Energy Devices

Conducted energy devices (CEDs) such as the Taser have garnered national—indeed, international—attention in the past few years. Police practitioners contend that CEDs are invaluable tools that minimize injuries to both officers and suspects during contentious confrontations, while critics argue that police may use CEDs in situations where such a high level of force is not warranted. Do police seem to be using CEDs appropriately? Gau, Mosher, and Pratt (2010) set out to address this question. They sought to determine whether suspects' race influenced the likelihood that police officers would deploy or threaten to deploy CEDs against those suspects. In an analysis of this sort, it is important to account for other variables that might be related to police use of CEDs or other types of force; therefore, the researchers included suspects' age, sex, and resistance level. They also measured officers' age, sex, and race. Finally, they included a variable indicating whether it was light or dark outside at the time of the encounter. The researchers found that police use of CEDs was driven primarily by the type and intensity of suspect resistance but that even controlling for resistance, Hispanic suspects faced an elevated probability of having CEDs either drawn or deployed against them.

▣ UNITS OF ANALYSIS

It seems rather self-evident, but nonetheless bears explicit mention, that every scientific study contains *something* that the researcher conducting the study gathers and examines. These "somethings" can be objects or entities such as rocks, people, molecules, or prisons. This is called the **unit of analysis** and it is, essentially, whatever the sample under study consists of. In criminal justice and criminology research, individual people are often the units of analysis. These individuals might be probationers, police officers, criminal defendants, or judges. Prisons, police departments, criminal incidents, or court records can also be units of analysis. Larger units are also popular—many studies focus on census tracks, block groups, cities, states, or even countries. Research Example 2.2 describes the methodological setup of a selection of criminal justice studies, each of which employed a different unit of analysis.

Unit of Analysis: The object or target of a research study.

RESEARCH EXAMPLE 2.2

Units of Analysis

Each of the following studies used a different unit of analysis.

1. *Does self-defensive gun use help or hurt crime victims?* Hart and Miethe (2009) used the National Crime Victimization Survey to cull all criminal victimizations in which the victim attempted to thwart the attack by using a firearm in self-defense. The units of analysis were, therefore, criminal incidents. The researchers' goal was to find

(Continued)

(Continued)

out whether self-defensive gun use by victims successfully warded off attack or, conversely, whether it made things worse. The statistical analyses revealed that although defensive gun use sometimes had adverse consequences for crime victims, this type of self-defense effort did prove advantageous for most victims under most circumstances.

2. *Is the individual choice to keep a firearm in the home affected by local levels of crime and police strength?* Kleck and Kovandzic (2009), using individual-level data from the General Social Survey (GSS) and city-level data from the FBI, set out to determine whether city-level homicide rates and the number of police per 100,000 city residents affected GSS respondents' likelihood of owning a firearm. There were two units of analysis in this study: individuals and cities. The statistical models indicated that high homicide rates and low police levels both modestly increased the likelihood that a given person would own a handgun; however, the relationship between city homicide rate and individual gun ownership decreased markedly when the authors controlled for whites' and other nonblacks' racist attitudes toward African Americans. It thus appeared that the homicide–gun ownership relationship was explained in part by the fact that those who harbored racist sentiments against blacks were more likely to own firearms regardless of the local homicide rate.

3. *Do neighborhoods within a single city have unique "criminal careers," and if so, what factors drive those careers?* Stults (2010) mapped out the trajectory of concentrated disadvantage and homicide over time in Chicago. He used Chicago homicide data broken down by neighborhood to analyze homicide trends in these neighborhoods from 1965 to 1995. The units of analysis were, therefore, neighborhoods. He analyzed the neighborhoods to determine whether area levels of structural disadvantage and social disorganization affected homicide rates over time. He found that, overall, the intensity of disadvantage and disorganization present in each neighborhood over time did impact neighborhood homicide rates.

▣ INDEPENDENT AND DEPENDENT VARIABLES

Researchers in criminal justice and criminology typically seek to examine relationships between two or more variables. Observed or **empirical** phenomena give rise to questions about the underlying forces driving those phenomena. One empirical event is homicide and city homicide rates. It is worthy of note, for instance, that Washington, D.C., has a higher violent crime rate than Portland, Oregon, has. Researchers usually want to do more than merely note empirical findings, however—they want to know *why* things are the way they are. They might, then, attempt to identify the criminogenic (crime-producing) factors that are present in Washington, D.C., but absent in Portland or, conversely, the protective factors possessed by Portland and lacked by Washington, D.C.

Researchers undertaking quantitative studies must specify **dependent variables** (DVs) and **independent variables** (IVs). Dependent variables are the empirical events that a researcher is attempting to explain. Neighborhood or city crime rates, ex-prisoner recidivism, and judicial sentencing decisions are examples of dependent variables. Independent variables are those factors that a researcher believes might either enhance or suppress the dependent variable. It might be predicted, for instance, that compared to their male counterparts, female judges sentence defendants less harshly because they tend to endorse the rehabilitation philosophy of punishment and the importance of keeping families and communities intact. The opposite, though, might also be predicted: An investigator may expect female judges to sentence offenders more harshly than do male judges pursuant to the hypothesis that women take special affront to criminal behaviors that threaten the safety and solidarity of the community. It is important to note that certain variables— crime rates, for instance—can be used as both independent and dependent variables across different studies. The designation of a certain phenomenon as an IV or DV depends upon the nature of the research study.

Empirical: Having the qualities of being measurable, observable, or tangible. Empirical phenomena are detectable with senses such as sight, hearing, or touch.

Dependent Variable: The phenomenon that a researcher wishes to study, explain, or predict.

Independent Variable: A factor or characteristic that is used to try to explain or predict a dependent variable.

▣ RELATIONSHIPS BETWEEN VARIABLES: A CAUTIONARY NOTE

It is vital to understand that *independent* and *dependent* are *not* synonymous with *cause* and *effect*, respectively. A particular independent variable might be related to a certain dependent variable, but this is far from definitive proof that the former is the cause of the latter. One problem is that there could be an **omitted variable** that explains the dependent variable as well as or even better than the independent variable does. The inadvertent exclusion of one or more important variables is called the **omitted variable bias**. The omitted variable bias is always a looming problem in criminal justice and criminology research because it is difficult to ensure that every influential independent variable is accounted for in a statistical model. Omitting IVs that exert significant influence on DVs can cause an IV that *is* included to appear to be strongly related to the DV when, in fact, the omitted variable is a better predictor. Research Example 2.3 offers an example of the omitted variable bias with respect to so-called crack babies.

> ## RESEARCH EXAMPLE 2.3
>
> ### The Problem of Omitted Variables
>
> Remember the "crack baby" panic? A media and political frenzy propelled this topic to the top of the national agenda for a time. The allegations were that "crack mothers" were abusing the drug while pregnant and were doing irreparable damage to their unborn children. Stories of low-birthweight, neurologically impaired newborns abounded. What the people caught up in this panic overlooked, though, was the fact that women who use crack cocaine while pregnant are also likely to use drugs such as tobacco and alcohol, which are known to cause problems to fetuses. They are also more likely to have low incomes and little or no access to prenatal health care. Finally, if a woman abuses crack—or any other drug—while pregnant, she may also be at risk for mistreating her child after its birth (see Logan, 1999, for a review). Given the array of variables that can affect a baby's development both in utero and after birth, do you think it is wise for the public to focus solely upon crack use? What public policy problems do you think could arise from such a focus?

Finally, independent variables cannot be viewed as the final determinants of dependent variables, because statistical analyses are examinations of aggregate trends. Uncovering an association between an IV and a DV means only that the presence of the IV has the tendency to either increase or reduce the DV in the sample as a whole—it is not by any means an indication that the IV-DV link holds true for every single person or object in the sample. For example, victims of early childhood trauma are more likely than nonvictims to develop substance abuse disorders later in life (see Dass-Brailsford & Myrick, 2010). Does this mean that every person who was victimized as a child has substance abuse problems as an adult? Certainly not! To draw such a conclusion would be to commit the **ecological fallacy**. Early trauma is a risk factor that may elevate the risk of later abuse of alcohol and other drugs, but it in no way guarantees that every abuse victim abuses or is addicted to drugs.

Omitted Variable: An independent variable that is significantly related to a dependent variable but has been erroneously excluded from the statistical analysis.

Omitted Variable Bias: The erroneous conclusion that there is a relationship between an independent and dependent variable when, in fact, that relationship is explained by a third variable that has not been included in the analysis.

Ecological Fallacy: The error of assuming that a statistical relationship that is present in a group applies uniformly to all individual people or objects within that group.

In sum, you should always be cautious when interpreting IV-DV relationships. It is better to think of IVs as *predictors* and DVs as *outcomes* rather than to view them as causes and effects. As the adage goes, correlation does not mean causation. Variables of all kinds may be related to each other, but it is important not to leap carelessly to causal conclusions on the basis of statistical associations.

▣ LEVELS OF MEASUREMENT

Every variable possesses a **level of measurement**. Levels of measurement are ways of classifying or describing variable type. There are two overarching classes of variables: **categorical** (also sometimes called *qualitative*) and **continuous** (also sometimes referred to as *quantitative*). Categorical variables are composed of groups or classifications that are represented with labels, while continuous variables are made up of numbers that measure how much of a particular characteristic a person or object possesses. Each of these variable types contains two subtypes. They are discussed in turn in the following sections.

Figure 2.1   Levels of Measurement

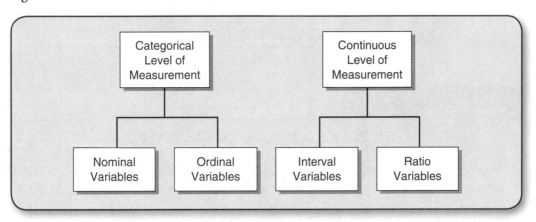

Level of Measurement: A variable's specific type or classification. There are four types: *nominal, ordinal, interval,* and *ratio*.

Categorical Variable: A variable that classifies people or objects into groups. There are two types: *nominal* and *ordinal*.

Continuous Variable: A variable that numerically measures the presence of a particular characteristic. There are two types: *interval* and *ratio*.

The Categorical Level of Measurement: Nominal and Ordinal Variables

Categorical variables are, not surprisingly, made up of categories. They represent ways of divvying people and objects up according to some characteristic. Categorical variables are subdivided into two types: *nominal* and *ordinal*. The **nominal level** of measurement is the most rudimentary of all the levels. It is the least descriptive and sometimes the least informative. Race is an example of a nominal-level variable. See Table 2.1 for an example (see also Data Sources 2.1 for a description of the data set used in Table 2.1). This variable is nominal because the different races and ethnicities are groups into which people are placed. The group labels offer descriptive information about the people or objects within each group.

Table 2.1   Race and Hispanic Origin of Stopped Drivers (PPCS)

Driver Race or Ethnicity	Frequency
White	9,054
Black/African American	1,076
Other non-Hispanic	1,135
Hispanic	578
	Total = 11,883

DATA SOURCES 2.1

THE POLICE-PUBLIC CONTACT SURVEY

The Bureau of Justice Statistics (BJS; see Data Sources 2.3 below) conducts the PPCS periodically as a supplement to the National Crime Victimization Survey (NCVS; see Data Sources 1.2). Interviews are conducted in English only. NCVS respondents aged 16 and older are asked about recent experiences they may have had with police. Variables include respondent demographics, the reason for respondents' most recent contact with police, whether the police used or threatened force against the respondents, the number of officers present at the scene, whether the police asked to search respondents' vehicles, and so on (BJS, 2005). This data set is used by BJS statisticians to estimate the total number of police-citizen contacts that take place each year, and it is used by researchers to study suspect, officer, and situational characteristics of police-public contacts.

These classifications, however, only represent *differences*; there is no way to arrange the categories in any meaningful rank or order. Nobody in one racial group can be said to have "more race" or "less race" than someone in another category—they are merely of different races. The same applies to gender. Most people identify as being either female or male, but members of one gender group do not have more or less gender relative to members of the other group.

Nominal Variable: A classification that places people or objects into different groups according to a particular characteristic that cannot be ranked in terms of quantity.

Ordinal Variable: A classification that places people or objects into different groups according to a particular characteristic that can be ranked in terms of quantity.

Ordinal variables are one step up from the nominal level in terms of descriptiveness because they can be ranked according to the quantity of a characteristic possessed by each person or object in a sample. University students' class level is an ordinal variable because freshmen, sophomores, juniors, and seniors can be rank-ordered according to how many credits they have earned. Numbers can also be represented as ordinal classifications when the numbers have been grouped into ranges like those in Table 2.2, where the income categories of respondents to the General Social Survey (GSS; see Data Sources 2.2) are shown.

Ordinal variables are useful because they allow people or objects to be ranked in a meaningful order. Ordinal variables are limited, though, by the fact that no algebraic techniques can be applied to them. This includes ordinal variables made from numbers such as those in Table 2.2. It is impossible, for instance, to subtract < *$1,000* from *$15,000 – $19,999*. It is likewise impossible to determine exactly how far apart two respondents are in their income levels. The difference between someone in the *$20,000 – $24,999* group and the ≥ *$25,000* group might only be one dollar if the former makes $24,999 and the latter makes $25,000. The difference could, though, be enormous if the person in the ≥ *$25,000* group has an annual family income of $500,000 per year. There is no way to figure this out from an assortment of categories like those in Table 2.2. The same limitation applies to the *year in college* variable—there is no way to find out how many more units a given senior has relative to any given junior.

Table 2.2   GSS Respondents' Reported Annual Family Income

Family Income	Frequency
< $1,000	20
$1,000 – $2,999	27
$3,000 – $3,999	23
$4,000 – $4,999	9
$5,000 – $5,999	15
$6,000 – $6,999	20
$7,000 – $7,999	28
$8,000 – $9,999	34
$10,000 – $14,999	105
$15,000 – $19,999	99
$20,000 – $24,999	130
≥ $25,000	1,264
	Total = 1,774

DATA SOURCES 2.2

THE GSS

The National Opinion Research Center (NORC) has conducted the General Social Survey (GSS) annually or every two years since 1972. Respondents are selected using a multistage clustering sample design. First, cities and counties are randomly selected. Second, block groups or districts are selected from those cities and counties. Trained researchers then canvass each block group or district on foot and interview people in person. Interviews are offered in both English and Spanish. The GSS contains a large number of variables. Some of these variables are asked in every wave of the survey, while others are only asked once. The variables include respondents' attitudes about religion, politics, abortion, the death penalty, gays and lesbians, persons of racial groups other than respondents' own, free speech, marijuana legalization, and a host of other topics (Davis & Smith, 2009).

The Continuous Level of Measurement: Interval and Ratio Variables

Continuous variables differ from categorical ones in that the former are represented not by categories but, rather, by numbers. **Interval-level variables** are numerical scales in which there are equal distances between all adjacent points on those scales. Ambient temperature is a classic example of an interval variable. This scale is measured using numbers representing degrees, and every point on the scale is exactly one degree away from the nearest points on each side. Twenty degrees Fahrenheit, for instance, is exactly 1 degree cooler than 21 degrees and exactly 4 degrees warmer than 16 degrees. *Age* is also interval level. Figure 2.2 shows the age of PPCS respondents who reported that their most recent encounter with police involved a threat of force.

Ratio variables are the other subtype within the continuous level of measurement. The ratio level resembles the interval level in that ratio, too, is numerical and has equal and known distance between adjacent points. The difference is that ratio-level scales have meaningful zero points that represent the absence of a given characteristic. Temperature, for instance, is not ratio level because the zeros in the various temperature scales are just placeholders. Zero does not signify an *absence* of temperature.

Table 2.3 is a frequency distribution displaying the number of state prisoners who were executed in 2010. These data come from the Bureau of Justice Statistics (BJS; see Data Sources 2.3). Can you explain why *number of persons executed* is a ratio-level variable?

Interval Variable: A quantitative variable that numerically measures the extent to which a particular characteristic is present or absent and does not have a true zero point.

Ratio Variable: A quantitative variable that numerically measures the extent to which a particular characteristic is present or absent and has a true zero point.

Figure 2.2 Age of PPCS Respondents Against Whom Police Threatened Force

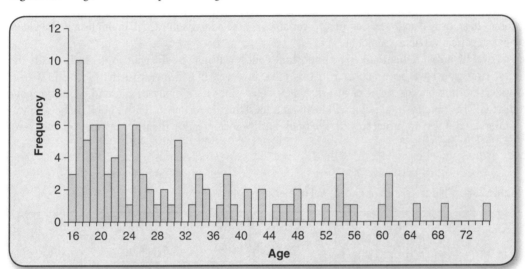

Table 2.3 Number of Persons Executed by States in 2010

Number Executed	f
0	38
1	5
2	1
3	3
5	1
8	1
17	1
	Total = 50

DATA SOURCES 2.3

THE BUREAU OF JUSTICE STATISTICS (BJS)

The Bureau of Justice Statistics (BJS) is the U.S. Department of Justice's repository for statistical information on criminal justice–related topics. BJS offers downloadable data and periodic reports on various topics that summarize the data and put them into a user-friendly format. Researchers, practitioners, and students all rely on BJS for accurate, timely information about crime, victims, sentences, prisoners, and more. Visit http://bjs.ojp.usdoj.gov/ and explore this valuable information source.

In the "real world" of statistical analysis, interval and ratio variables are generally used interchangeably. It is the overarching categorical-versus-continuous distinction that usually matters most when it comes to statistical analyses. Continuous variables can be added, subtracted, multiplied, and divided, but categorical variables cannot be.

Level of measurement is a very important concept. It may be difficult to grasp if this is the first time you have been exposed to this idea; however, it is imperative that you gain a firm understanding because level of measurement determines what analyses can and cannot be conducted. This fundamental point will form an underlying theme of this entire book, so be sure you understand it. Do not proceed with the book until you can readily identify a given variable's level of measurement.

Table 2.4   Characteristics of Each Level of Measurement

Level of Measurement	Variable Characteristic		
	Rank-orderable	Equal intervals	True zero
Nominal			
Ordinal	✓		
Interval	✓	✓	
Ratio	✓	✓	✓

STUDY TIP: SELECTING THE BEST LEVEL OF MEASUREMENT

When you have a choice about level of measurement, use the highest level possible. You can always make a continuous variable categorical later after the data have been collected, but you can never make a categorical variable continuous.

CHAPTER SUMMARY

This chapter discussed the concept of a variable. You also read about units of analysis, independent variables, dependent variables, and two of the problems—the omitted variable bias and the ecological fallacy—that require caution to be taken in the interpretation of IV-DV relationships. It is important to view statistical associations as relationships rather than strict cause-effect sequences.

This chapter also described the two primary levels of measurement: categorical and continuous. Categorical variables are qualitative groupings or classifications into which people or objects are placed on the basis of some characteristic. The two subtypes of categorical variables are nominal and ordinal. These two kinds of variables are quite similar in appearance, with the distinguishing feature being that nominal variables cannot be rank-ordered, while ordinal variables can be. Continuous variables are quantitative measurements of the presence or absence of a certain characteristic in a group of people or objects. Interval and ratio variables are both continuous. The difference between them is that ratio-level variables possess true zero points and interval-level variables do not.

You must understand this concept and be able to identify the level of measurement of any given variable because in statistics, the level at which a variable is measured is one of the biggest determinants of the analytic techniques that can be employed. In other words, each type of statistical analysis can be used with some levels of measurement and cannot be used with others. Employing the wrong statistical procedure can produce wildly inaccurate results and conclusions. Even the relatively simple procedures such as constructing graphs (Chapter 3) are affected by levels of measurement. You must therefore possess an understanding of level of measurement before leaving this chapter.

KEY TERMS

Variable	Omitted variable	Continuous variable
Constant	Omitted variable bias	Nominal variable
Unit of analysis	Ecological fallacy	Ordinal variable
Empirical	Level of measurement	Interval variable
Dependent variable	Categorical variable	Ratio variable
Independent variable		

CHAPTER 2 REVIEW PROBLEMS

1. A researcher wishes to test the hypothesis that low education affects crime. She gathers a sample of people ages 25 and over.
   a. What is the independent variable?
   b. What is the dependent variable?
   c. What is the unit of analysis?

2. A researcher wishes to test the hypothesis that arrest deters recidivism. She gathers a sample of people who have been arrested.
   a. What is the independent variable?
   b. What is the dependent variable?
   c. What is the unit of analysis?

3. A researcher wishes to test the hypothesis that poverty affects violent crime. He gathers a sample of neighborhoods.
   a. What is the independent variable?
   b. What is the dependent variable?
   c. What is the unit of analysis?

4. A researcher wishes to test the hypothesis that prison architectural design affects the number of inmate-on-inmate assaults that take place inside a facility. He gathers a sample of prisons.

    a. What is the independent variable?

    b. What is the dependent variable?

    c. What is the unit of analysis?

5. A researcher wishes to test the hypothesis that the amount of money a country spends on education, health, and welfare affects the level of violent crime in that country. She gathers a sample of countries.

    a. What is the independent variable?

    b. What is the dependent variable?

    c. What is the unit of analysis?

6. A researcher wishes to test the hypothesis that police officers' job satisfaction affects the length of time they stay in their jobs. He gathers a sample of police officers.

    a. What is the independent variable?

    b. What is the dependent variable?

    c. What is the unit of analysis?

7. A researcher wishes to test the hypothesis that the location of a police department in either a rural or an urban area affects starting pay for entry-level police officers. She gathers a sample of police departments.

    a. What is the independent variable?

    b. What is the dependent variable?

    c. What is the unit of analysis?

8. A researcher wishes to test the hypothesis that the level of urbanization in a city or town affects residents' social cohesion. She gathers a sample of municipal jurisdictions (cities and towns).

    a. What is the independent variable?

    b. What is the dependent variable?

    c. What is the unit of analysis?

9. Explain the omitted variable bias.

10. Explain the ecological fallacy.

11. Identify the level of measurement of each of the following variables.

    a. Suspects' race measured as *white, black, Latino,* and *other.*

    b. The age at which an offender was arrested for the first time.

    c. The sentences received by convicted defendants, measured as *jail, prison, probation, fine,* and *other.*

    d. The total number of status offenses that adult offenders reported having committed as juveniles.

    e. The amount of money, in dollars, that a police department collects annually from drug asset forfeitures.

    f. Prison security level, measured as *minimum, medium,* and *maximum.*

    g. Trial judges' gender.

12. Identify the level of measurement of each of the following variables.

    a. The amount of resistance a suspect displays toward the police, measured as *not resistant, somewhat resistant,* or *very resistant.*

    b. The number of times someone has shoplifted in her or his life.

c. The number of times someone has shoplifted, measured as *0 – 2, 3 – 5,* or *6 or more.*
d. The type of attorney a criminal defendant has at trial, measured as *privately retained* or *publicly funded.*
e. In a sample of juvenile delinquents, whether or not those juveniles have substance abuse disorders.
f. Prosecutors' charging decisions, measured as *filed charges* and *did not file charges.*
g. In a sample of offenders sentenced to prison, the number of days in their sentences.

13. If a researcher is conducting a survey and wants to ask respondents about their self-reported involvement in shoplifting, there are a few different ways he could phrase this question.

   a. Identify the level of measurement that each type of phrasing shown below would produce.
   b. Explain which of the three possible phrasings would be the best one to choose and why this is.

   Possible phrasing 1: *How many times have you taken small items from stores without paying for those items?*

   Please write in: _____

   Possible phrasing 2: *How many times have you taken small items from stores without paying for those items? Please circle one of the following:*

   Never   1 – 2 times   3 – 4 times   5+ times

   Possible phrasing 3: *Have you ever taken small items from stores without paying for those items? Please circle one of the following:*

   Yes   No

14. If a researcher is conducting a survey and wants to ask respondents about the number of times each of them has been arrested in his or her life, there are a few different ways she could phrase this question. Write down the three possible phrasing methods.

15. The following table contains BJS data on the number of prisoners under sentence of death in 2008, by region. Use the table to do the following.

Number of Prisoners Under Sentence of Death at Year-End 2008 (BJS)

Region	Prisoners
Northeast	234
Midwest	270
South	1,706
West	946
	Total = 3,156

   a. Identify the level of measurement of the variable *region.*
   b. Identify the level of measurement of *number of prisoners under sentence of death.*

16. The following table contains NCVS data showing the number of victimized respondents who reported their victimization to police. The data are broken down by respondents' household income level. Use the table to do the following.

Number of NCVS Respondents Who Reported Their
Victimization to Police, by Household Income

Income	Reported Victimization?		
	Yes	No	Total
$12,499 or less	670	987	1,657
$12,500 – $24,999	857	1,195	2,052
$25,000 – $49,999	1,481	2,022	3,503
$50,000 or more	1,986	3,104	5,090
Total	4,994	7,308	$N = 12,302$

a. Identify the level of measurement of the *variable income.*
b. Identify the level of measurement of the variable *reported victimization?*

17. Haynes (2011) conducted an analysis to determine whether victim advocacy affects offender sentencing. She measured victim advocacy as a yes/no variable indicating whether or not there was a victim witness office located inside the courthouse. She measured sentencing as the number of months of incarceration imposed on convicted offenders.

a. Identify the independent variable in this study.
b. Identify the level of measurement of the independent variable.
c. Identify the dependent variable in this study.
d. Identify the level of measurement of the dependent variable.

18. Bouffard and Piquero (2010) wanted to know whether arrested suspects' perceptions of the way police treated them during the encounter affected the likelihood that those suspects would commit more crimes in the future. Their sample consisted of males who had been arrested at least once during their lives. They measured suspects' perceptions of police behavior as *fair* or *unfair*. They measured recidivism as the number of times suspects came into contact with police after that initial arrest.

a. Identify the independent variable in this study.
b. Identify the level of measurement of the independent variable.
c. Identify the dependent variable in this study.
d. Identify the level of measurement of the dependent variable.

19. Kleck and Kovandzic (2009; see Research Example 2.2) examined whether the level of homicide in a particular city affected the likelihood that people in that city would own firearms. They measured homicide as the number of homicides that took place in the city in one year divided by the total city

population. They measured handgun ownership as whether survey respondents said they did or did not own a gun.

  a. Identify the independent variable used in this study.
  b. Identify the level of measurement of the independent variable.
  c. Identify the dependent variable in this study.
  d. Identify the level of measurement of the dependent variable.

20. Gau, Mosher, and Pratt (2010; see Research Example 2.1) examined whether suspects' race or ethnicity influenced the likelihood that police would brandish or deploy Tasers against them. They measured race as *white, Hispanic, black,* or *other.* They measured Taser usage as *Taser used* or *some other type of force used.*

  a. Identify the independent variable used in this study.
  b. Identify the level of measurement of the independent variable.
  c. Identify the dependent variable in this study.
  d. Identify the level of measurement of the dependent variable.

REFERENCES

Bouffard, L. A., & Piquero, N. L. (2010). Defiance theory and life course explanations of persistent offending. *Crime & Delinquency, 56*(2), 227–252.

Bureau of Justice Statistics. (2005). *The police-public contact survey, 2005.* Ann Arbor, MI: Inter-University Consortium for Political and Social Research.

Dass-Brailsford, P., & Myrick, A. C. (2010). Psychological trauma and substance abuse: The need for an integrated approach. *Trauma, Violence, & Abuse, 11*(4), 202–213.

Davis, J. A., & Smith, T. W. (2009). *General social surveys, 1972–2008.* Chicago: National Opinion Research Center, producer, 2005; Storrs, CT: Roper Center for Public Opinion Research, University of Connecticut.

Gau, J. M., Mosher, C., & Pratt, T. C. (2010). An inquiry into the impact of suspect race on police use of Tasers. *Police Quarterly, 13*(1), 27–48.

Hart, T. C., & Miethe, T. D. (2009). Self-defensive gun use by crime victims: A conjunctive analysis of its situational contexts. *Journal of Contemporary Criminal Justice, 25*(1), 6–19.

Haynes, S. H. (2011). The effects of victim-related contextual factors on the criminal justice system. *Crime & Delinquency, 57*(2), 298–328.

Kleck, G., & Kovandzic, T. (2009). City-level characteristics and individual handgun ownership: Effects of collective security and homicide. *Journal of Contemporary Criminal Justice, 25*(1), 45–66.

Logan, E. (1999). The wrong race, committing crime, doing drugs, and maladjusted for motherhood: The nation's fury over "crack babies." *Social Justice, 26*(1), 115–138.

Stults, B. J. (2010). Determinants of Chicago neighborhood homicide trajectories: 1965–1995. *Homicide Studies, 14*(3), 244–267.

Organizing, Displaying, and Presenting Data

Data are usually stored in electronic files for use with software programs designed to conduct statistical analyses. The program SPSS is one of the most common data software programs in criminal justice and criminology. A typical SPSS file may look like this (see Figure 3.1):

Figure 3.1   SPSS Data File

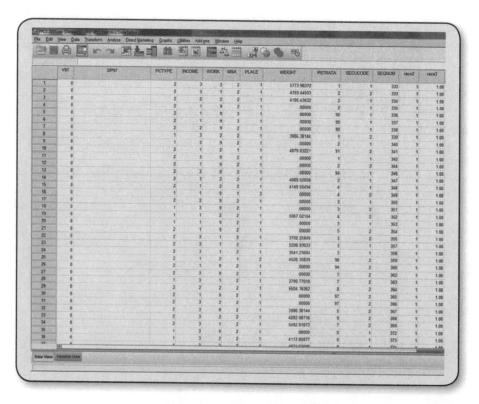

The data in Figure 3.1 are from the Bureau of Justice Statistics' 2005 Police-Public Contact Survey (PPCS; see Data Sources 2.1). Each row (horizontal line) in the grid represents one respondent, and each column (vertical line) represents one variable. Where any given row and column meet is a **cell** containing a given person's response to a particular question.

Your thoughts as you gaze upon the data screen in Figure 3.1 can probably be aptly summarized as "Huh?" That is a very appropriate response, because there is no way to make sense of the data when they are in this raw format. This brings us to the topic of this chapter: methods for organizing, displaying, and presenting data. You can see from the figure that something has to be done to the data set to get it into a useful format. This chapter will teach you how to do just that.

Cell: The place in a table where a row and column meet.

Chapter 2 introduced you to levels of measurement (nominal, ordinal, interval, and ratio) and explained that a variable's level of measurement determines which statistical techniques are and are not appropriate for that variable. Graphs and charts are no exception. There exist a variety of data displays, and many of them can only be used with variables of particular types. As you read this chapter, take notes on two main concepts: (1) the proper construction of each type of data display and (2) the levels of measurement for which each display type is applicable.

▣ DATA DISTRIBUTIONS

Univariate Displays: Frequencies, Proportions, and Percentages

Perhaps the most straightforward type of pictorial display is the **univariate** (one variable) **frequency** distribution. A frequency is simply a raw count; it is the number of times a particular characteristic appears in a data set. A frequency distribution is a tabular display of frequencies. Table 3.1 shows the frequency distribution for the variable *respondent gender* in the 2005 PPCS survey. Let us pause and consider two new symbols. The first is the *f* that sits atop the right-hand column in Table 3.1. This stands for *frequency*. You can see that there are 38,078 males in this sample. An alternative way of phrasing this is that the characteristic *male* occurs 38,078 times. A more formal way to write this would be $f_{male} = 38,078$; for females, $f_{female} = 42,159$. The second new symbol is the *N* found in the bottom right-hand cell. This represents the total sample size; here, $f_{male} + f_{female} = N$. Numerically, 38,078 + 42,159 = 80,237.

Table 3.1   Gender of PPCS Respondents

Gender	f
Male	38,078
Female	42,159
	N = 80,237

Univariate: Involving one variable.

Frequency: A raw count of the number of times a particular characteristic appears in a data set.

Proportion: A standardized form of a frequency that ranges from 0.00 to 1.00.

Percentage: A standardized form of a frequency that ranges from 0.00 to 100.00.

Raw frequencies are of limited use in graphical displays because they are often difficult to interpret and do not offer much information about the variable being examined. What is needed is a way to standardize the numbers to enhance interpretability. **Proportions** do this. Proportions are defined as the number of times a particular characteristic appears in a sample relative to the total sample size. Formulaically:

$$p = \frac{f}{N}, \text{where}$$

Formula 3(1)

p = proportion

f = raw frequency

N = total sample size.

Proportions range from 0.00 to 1.00. A proportion of exactly 0.00 indicates a complete absence of a given characteristic. If there were no males in the PPCS, their proportion would be $p_{male} = \frac{0}{80,237} = 0.00$. Conversely, a trait with a proportion of 1.00 would be the only characteristic present in the sample. If the PPCS contained only men, then the proportion of the sample that was male would be $p_{male} = \frac{80,237}{80,237} = 1.00$.

Another useful technique is to convert frequencies into **percentages** (abbreviated pct). Percentages are a variation on proportions and convey the same information, but percentages offer the advantage of being more readily interpretable by the general public. Percentages are computed similarly to proportions, with the added step of multiplying by 100:

$$pct = \left(\frac{f}{N}\right)100.$$

Formula 3(2)

Proportions and percentages can be added to frequencies to form a fuller, more informative display like that in Table 3.2. Note that the two proportions (p_{males} and $p_{females}$) sum to 1.00 because the two categories contain all respondents in the sample; the percentage column sums to 100.00 for the same reason. The Σ symbol in the table is the Greek letter *sigma* and is a summation sign. When all cases in a sample have been counted once and only once, proportions will sum to 1.00 and percentages to 100.00 or within rounding error of these totals.

Table 3.2 Gender of PPCS Respondents

Sex	f	p	pct
Male	38,078	.47	47.46
Female	42,159	.53	52.54
	$N = 80,237$	$\Sigma = 1.00$	$\Sigma = 100.00$

Another useful technique for frequency distributions is the computation of **cumulative** measures. Cumulative frequencies, cumulative proportions, and cumulative percentages can facilitate meaningful interpretation of distributions, especially when data are continuous. Consider Table 3.3 below, which contains the ratio-level variable measuring the number of criminal incidents experienced by married male respondents to the National Crime Victimization Survey (NCVS; see Data Sources 1.2) in the six months leading up to the interview. The number of incidents reported and the frequency of each response are located in the two leftmost columns. To their right is a column labeled *cf*, which stands for *cumulative frequency*. The *cp* and *cpct* columns contain cumulative proportions and percentages, respectively.

Table 3.3 Number of Criminal Victimizations Reported by Married, Male Respondents to the 2004 NCVS

Victimizations	f	cf	p	cp	pct	cpct
0	5	5	.02	.02	2.28	2.28
1	144	149	.66	.68	65.75	68.03
2	49	198	.22	.90	22.37	90.40
3	14	212	.06	.96	6.39	96.79
4	2	214	.01	.97	0.91	97.70
5	5	219	.02	.99	2.28	99.98
	$N = 219$		$\Sigma = .99$		$\Sigma = 99.98$	

Cumulative columns are constructed by summing the *f*, *p*, and *pct* columns successively from row to row. In Table 3.3's *cf* column, for instance, 149 is the sum of the two preceding numbers, 5 and 144. In the *cp* column, likewise, .68 = .02 + .66. Cumulatives allow for assessments of whether the data are clustered at one end of the scale or spread fairly equally throughout. In Table 3.3, it can be readily concluded that the data cluster at the low end of the scale, as .68 (or 68.03%) of respondents reported one or no victimizations, and .90 (or 90.40%) reported two or fewer.

Cumulative: A frequency, proportion, or percentage obtained by adding a given number to all numbers below it.

Univariate Displays: Rates

Suppose someone informed you that 2,199,125 burglaries were reported to the police in 2009. What would you make of these numbers? Nothing, probably, because raw numbers of this sort are simply not very useful. They lack a vital component—a denominator. The question that would leap to your mind immediately is "2,199,125 *out of what?*" You would want to know if these numbers were derived from a single city, from a single state, or from the United States as a whole. This is where rates come in. A *rate* is a method of standardization that involves dividing the number of events of interest (for example, burglaries) by the total population:

$$Rate = \frac{f}{population}.$$

Formula 3(3)

Table 3.4 contains 2009 UCR data (see Data Sources 1.1) for the property crime index offense categories. The column titled *Rate* displays the rate per capita that is obtained by employing Formula 3(3).

Table 3.4   UCR Index Offenses and Offense Rates, 2009

Crime	f	Rate	Rate per 10,000
Burglary	2,199,125	.01	71.63
Larceny	6,327,230	.02	206.09
Motor vehicle theft	794,616	.003	25.88

Population = 307,006,550

Note how tiny the numbers in the rate column are. Rates per capita do not make sense in the context of low-frequency events like crime because they end up being so small. It is, therefore, customary to multiply rates by a certain factor. This factor is usually 1,000, 10,000, or 100,000. You should select the multiplier that makes the most sense with the data. In Table 3.4, the 10,000 multiplier has been used to form the *Rate per 10,000* column. Multiplying in this fashion lends clarity to rates because now it is no longer the number of crimes *per person* but, rather, the number of crimes *per 10,000 people*. If you randomly selected a sample of 10,000 people from the general population, you would expect 71.63 of them to have been the victim of a burglary in the past year and 206.09 of them to have experienced larceny. These numbers and their interpretation are more real and more tangible than those derived using Formula 3(3) without a multiplier.

Bivariate Displays: Contingency Tables

Researchers are often interested not just in the frequency distribution of a single variable (a univariate display) but, rather, in the overlap between two variables. The PPCS asks respondents whether, during their most recent traffic stop, the officer or officers asked permission to search their vehicle. Respondents are also asked to report their gender. The *asked to search* variable can be added to the respondent gender variable to form a **contingency table** (also sometimes called *crosstabs*). This is a **bivariate** display, meaning it contains two variables.

Contingency table: A table showing the overlap between two variables.

Bivariate: Analysis involving two variables. Usually, one is designated the independent variable and the other the dependent variable.

Table 3.5 PPCS Respondent Gender and Police Request to Search Vehicle During Traffic Stop (Frequencies)

	Police Asked to Search?		
Gender	Yes	No	Row Total
Male	104	2,558	2,662
Female	18	1,886	1,904
Column Total	122	4,444	N = 4,566

Researchers studying police behavior might be curious as to whether male drivers are more or less likely than female drivers to be asked to consent to a vehicle search during a traffic stop. The raw frequencies offer no information to this effect, but proportions and percentages can be used to shed some light on the matter. Note that there are two types of proportions and percentages that can be computed in a bivariate contingency table: *row* and *column*. Row proportions and percentages are computed using the row totals in the denominator, while column proportions and percentages employ the column totals. If we want to discover the percentage of persons of each gender who were subject to search requests, row totals are the appropriate denominators. Column marginals would be used if the question under investigation were, "Among those drivers asked for consent to search, what percentage was male and what percentage was female?" Table 3.6 shows the percentage distribution.

Two conclusions can be drawn from this table. First, the vast majority of drivers of both sexes were *not* asked for consent to search. Second, a larger percentage of males were asked relative to females; nearly

Table 3.6 PPCS Respondent Gender and Police Request to Search Vehicle During Traffic Stop (Row Percentages)

	Police Asked to Search?		
Gender	Yes	No	Row Total
Male	$\left(\frac{104}{2662}\right)100 = 3.91$	$\left(\frac{2558}{2662}\right)100 = 96.09$	100.00
Female	$\left(\frac{18}{1904}\right)100 = .95$	$\left(\frac{1886}{1904}\right)100 = 99.06$	100.01
			N = 4,566

4 % of males but less than 1 % of females were asked. It thus appears that male stopped drivers are more likely than female stopped drivers to be the recipients of police officers' request to search their vehicles.

▣ GRAPHS AND CHARTS

Frequency, proportion, and percentage distributions are helpful ways of summarizing data; however, they are rather dull to look at. It is sometimes desirable to arrange data in a more attractive format. If you were giving a presentation to a local police department or district attorney's office, for instance, you would not want to throw numbers at your audience for 20 or 30 minutes. The monotony would bore them to pieces. Presentations can be diversified by the introduction of charts and graphs, of which there are many different types. This chapter will concentrate on five of the most common: *pie charts, bar graphs, histograms, frequency polygons,* and *line graphs.*

Categorical Variables: Pie Charts

Pie charts can only be used with categorical data, and they are most appropriate for variables that have relatively few **classes** (that is, categories or groups) because pie charts get messy fast. A good general rule is to use a pie chart only when a variable contains five or fewer classes. Pie charts are based on percentages; the entire circle represents 100% and the "slices" are sized according to their level of contribution to that total.

The variable that will be used here to illustrate a pie chart is the race and ethnicity of stopped drivers from Table 2.1. The first step is to transform the raw frequencies into percentages using Formula 3(2). Once percentages have been computed, the pie chart can be built by dividing 100% into its constituent parts. Figure 3.2 contains the pie chart. Flip back to Table 2.1 and compare this pie chart to the raw frequency distribution to note the dramatic difference between the two presentation methods.

Figure 3.2   Race and Ethnicity of Stopped Drivers

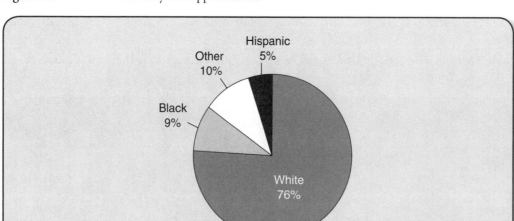

Classes: The categories or groups within a nominal or ordinal variable.

Categorical Variables: Bar Graphs

Like pie charts, bar graphs are meant to be used with categorical data; unlike pie charts, though, bar graphs can accommodate variables with many classes without damage to the charts' readability. Bar graphs are thus more flexible than pie charts are. For variables with five or fewer classes, pie charts and bar graphs may be equally appropriate; when there are six or more classes, bar graphs should be used.

In Chapter 1, you learned that one of the reasons for the discrepancy between crime prevalence as reported by the Uniform Crime Reports (UCR) and the National Crime Victimization Survey (NCVS) is that a substantial portion of crime victims do not report the incident to police. Hart (2003) analyzed NCVS data from the year 2000 and reported the percentage of people victimized by different crime types who reported their victimization to police. Figure 3.3 contains a bar graph illustrating the percentage of victims who contacted the police. Bar graphs provide ease of visualization and interpretation. It is simple to see from Figure 3.3 that substantial portions of all types of victimizations are not reported to the police, and motor vehicle theft is the most reliably reported crime.

Figure 3.3 Percent of Victimizations Reported to Police

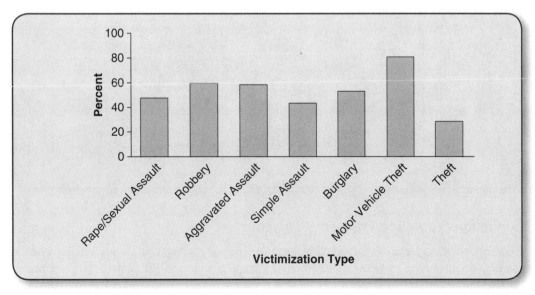

Rates can also be presented as bar graphs. Figure 3.4 is a bar graph of the rates in Table 3.4.

Figure 3.4   2009 Property Crime Rates per 10,000

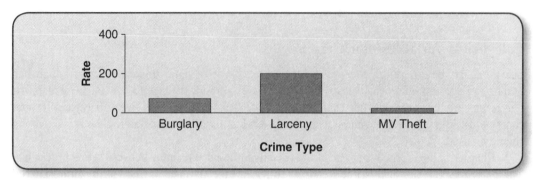

Continuous Variables: Histograms

Histograms are for use with continuous data. Histograms resemble bar charts, with the exception that in histograms, the bars touch one another. In bar charts, the separation of the bars signals that each category is distinct from the others; in histograms, the absence of space symbolizes the underlying continuous nature of the data. Figure 3.5 contains a histogram for the variable *age at sentencing* from the 2004 National Judicial Reporting Program (NJRP; see Data Sources 3.1). The sample includes persons who were convicted of murder and sentenced to prison. A histogram is appropriate for the variable *age* because, although it is customary to measure age in years, the passage of time can be broken down into months, days, minutes, seconds, and so on. Age, as a measure of time, is continuous and must be represented using a histogram.

Figure 3.5   Age at Sentencing Among Persons Sentenced to Prison for Murder

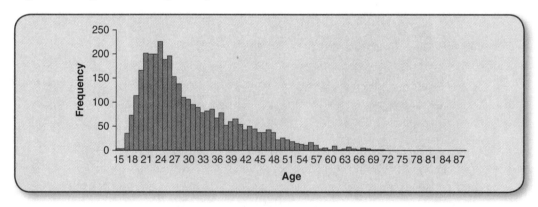

Continuous Variables: Frequency Polygons

Frequency polygons are an alternative to histograms. There is no "right" or "wrong" choice when it comes to deciding whether to use a histogram or a frequency polygon with a particular continuous variable; the best strategy is to mix it up a bit so that you are not using the same chart type repeatedly. Figure 3.6 contains the frequency polygon for the NJRP *age at sentencing* variable used in Figure 3.5.

Frequency polygons are created by placing a dot in the places where the tops of the bars would be in a histogram and then connecting those dots with a line.

Figure 3.6   Age at Sentencing Among Persons Sentenced to Prison for Murder

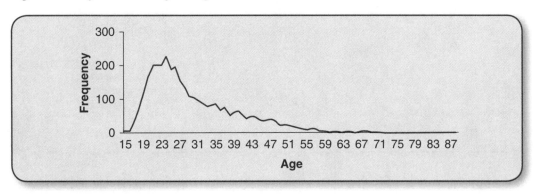

DATA SOURCES 3.1

THE NJRP

Every two years, the Bureau of Justice Statistics (BJS) randomly selects 300 counties using sampling methods that ensure a sample that is representative of counties nationwide. The data for the NJRP are extracted from official court records and include information about the offense (for example, offense type and number of offenses), the offender (for instance, age, sex, and race), and sentence (such as type of sentence imposed and length of sentence).

Longitudinal Variables: Line Charts

People who work with criminal justice and criminology data often encounter **longitudinal** variables. Longitudinal variables are measured repeatedly over time. Crime rates are often presented longitudinally as a means of determining **trends**. Line graphs can make discerning trends easy.

Longitudinal variables: Variables measured repeatedly over time.

Trends: Patterns that indicate whether something is increasing, decreasing, or staying the same over time.

Figure 3.7 contains a line graph of UCR data measuring the annual percentage of hate crime victims that were targeted on the basis of their sexual orientation from 1996 to 2008. It appears from this graph

Figure 3.7   Percent of Hate Crime Victims Targeted on the Basis of Sexual Orientation, 1996–2008

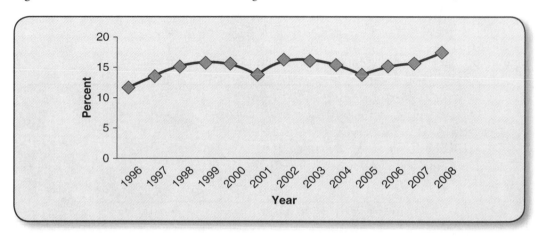

that the percentage of all hate crimes that are motivated by sexual orientation bias has fluctuated over the years, but there has been a general upward trend since 2005.

▣ GROUPED DATA

The overarching purpose of a frequency distribution, chart, or graph is to display data in an accessible, readily understandable format. Sometimes, though, continuous variables do not lend themselves to tidy displays. Consider Table 3.7's frequency distribution for the amount of money, in dollars per person, that local governments in each state spent on criminal justice operations (Morgan, Morgan, & Boba, 2010; Data Sources 3.2). Figure 3.7 displays a histogram of the data. You can see that neither the frequency distribution nor the histogram are useful; there are too many values, and most of the values occur only once in the data set. There is no way to discern patterns or draw any meaningful conclusion from these data displays.

Table 3.7   Per Capita Local Government Expenditures on Criminal Justice, Ungrouped

Dollars Per Capita	f	Dollars Per Capita	f	Dollars Per Capita	f
123	1	266	1	357	2
164	1	267	2	358	1
169	1	268	1	374	1
188	1	272	1	375	1
209	1	273	1	387	1
214	1	281	1	390	1
215	2	287	1	414	1

(Continued)

(Continued)

Dollars Per Capita	f	Dollars Per Capita	f	Dollars Per Capita	f
220	1	292	1	427	1
230	1	306	2	433	1
246	1	310	1	439	1
248	1	321	1	462	1
251	1	345	1	500	1
252	1	348	1	552	1
255	1	351	1	603	1
258	1	354	1	623	1
263	1			N = 50	

DATA SOURCES 3.2

CQ PRESS' STATE FACTFINDER SERIES

The Factfinder Series' Crime Rankings are compilations of various crime and crime-related statistics from the state and local levels. These volumes are comprehensive reports containing data derived from the FBI, Bureau of Justice Statistics, Census Bureau, and Drug Enforcement Administration. The data used here come from Morgan, Morgan, and Boba (2010).

Grouping the data can provide a solution to this problem by transforming a continuous variable (either interval or ratio) into an ordinal one. There are several steps to grouping. First, find the range in the data by subtracting the smallest number from the highest. Second, select the number of intervals you want to use. This step is more art than science; it might take you a bit of trial and error to determine the number of intervals that is best for your data.

The ultimate goal is to find a middle ground between having too few and too many intervals—too few can leave your data display flat and uninformative, while too many will defeat the whole purpose of grouping.

Third, determine interval width by dividing the range by the number of intervals. This step is probably best illustrated in formulaic terms:

$$Width = \frac{Range}{Intervals}.$$

This will often produce a number with a decimal, so round either up or down depending on your reasoned judgment as to the optimum interval width for your data. Fourth, construct the stated class limits by starting with the smallest number in the data set and creating intervals of the width determined in Step 3 until you run out of numbers. Finally, make a new frequency (*f*) column by counting the number of people or objects within each stated class interval.

Let us group the legal expenditure data in Table 3.7. First, we need the range:

$$Range = 623 - 123 = 500.$$

Now, we have to choose the number of intervals we want to use. With a large range like 500, it is advisable to select relatively few intervals so that each interval will encompass enough raw values to make it meaningful. We will start with 10 intervals. The next step is to compute the interval width. Using the formula from above—

$$Width = \frac{500}{10} = 50$$

—each interval will contain 50 raw scores. Now the stated class limits can be constructed. See the left-hand column in Table 3.8 below. There are three main points to keep in mind when building stated class limits. The stated limits must be *inclusive*—in this example, the first interval contains the number 123, the number 172, and everything in between; *mutually exclusive*; and *exhaustive*. Once the stated class limits have been determined, the frequency for each interval is calculated by summing the number of raw data points that fall into each stated class interval. The sum of the frequency column in a grouped distribution should equal the sum of the frequencies in the ungrouped distribution.

You can see that Table 3.8 is much neater and more concise than Table 3.7. It is more condensed and easier to read. Where you will really see the difference, though, is in the histogram. Take a look at Figure 3.8 and compare it to Figure 3.7. Quite an improvement! It has a real shape now. This demonstrates the utility of data grouping.

Table 3.8 Per Capita Local Government Expenditures on Criminal Justice, Grouped

Stated Class Limits	f
123 – 172	3
173 – 222	6
223 – 272	13
273 – 322	8
323 – 372	7
373 – 422	5
423 – 472	4
473 – 522	1
523 – 572	1
573 – 622	1
623 – 672	1
	N = 50

Figure 3.8 Per Capita Local Government Expenditures on Criminal Justice, Grouped

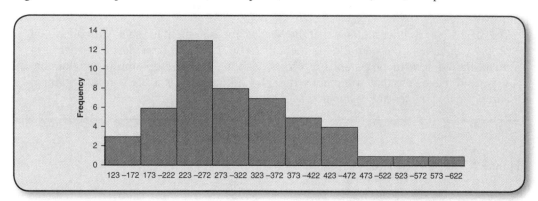

▣ SPSS

To obtain a frequency distribution, click on *Analyze* → *Descriptive Statistics* → *Frequencies,* as shown in Figure 3.9 on the next page. Select the variable you want from the list on the left side and either drag it to the right or click the arrow to move it over. For this illustration, we will use the *offense5* variable from the National Judicial Reporting Program (NJRP) data set. This variable groups defendants' conviction charges into five categories.

SPSS 3.1   GIGO ALERT!

Using the wrong statistical technique will produce unreliable and potentially misleading results. Statistical software programs will generally not alert you to errors of this sort; they will give you output even if that output is garbage. It is your responsibility to ensure that you are using the program correctly.

SPSS 3.2   THE SPSS FILE EXTENSION

The *.sav* extension signifies an SPSS data file; you should memorize this so that anytime you see this extension, you recognize it immediately. SPSS is the only program that will open a file with the *.sav* extension, so make sure you are working on a computer equipped with SPSS when you sit down to do homework or practice problems involving *.sav* files.

Selecting *OK* will produce the output shown in Figure 3.10.

The SPSS Chart Builder (accessible from the *Graphs* drop-down menu) allows you to select a chart type and choose the variable you want to use. We will use *offense5* again. In Figure 3.11, the chosen graph is a pie chart, which means that the *Count* in the *Element Properties* box must be changed to *Percent*. Clicking *Apply* and *OK* produces the chart, as shown in Figure 3.12.

SPSS 3.3   CHART BUILDER TIP

The SPSS Chart Builder requires that the level of measurement for each variable be set properly. SPSS will not permit certain charts to be used with some levels of measurement. Before constructing graphs or charts, visit the *Measure* column in the Variable View and make sure that continuous variables are marked as *Scale* and that nominal and ordinal variables are designated as such.

Figure 3.9   Running Frequencies in SPSS

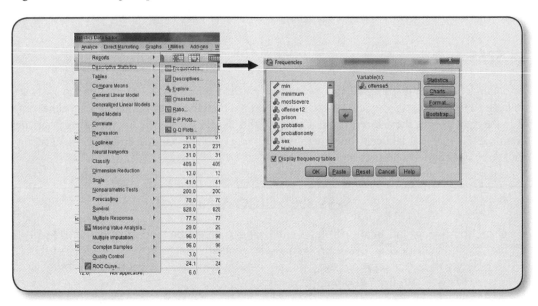

Figure 3.10   SPSS Frequency Output

MOST SERIOUS OFFENSE: 5 CATEGORIES					
		Frequency	Percent	Valid Percent	Cumulative Percent
Valid	Violent offenses	86184	18.3	18.3	18.3
	Property offenses	128958	27.3	27.3	45.6
	Drug offenses	168963	35.8	35.8	81.4
	Weapon offenses	18293	3.9	3.9	85.3
	Other offenses	69247	14.7	14.7	100.0
	Total	471645	100.0	100.0	

Figure 3.11   Using the Chart Builder to Create a Pie Chart

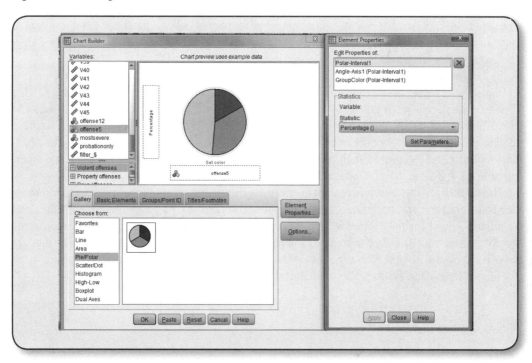

Figure 3.12   SPSS Pie Chart

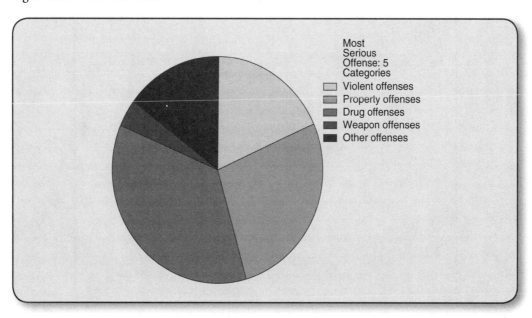

SPSS can also be used to transform raw numbers into rates. To demonstrate this, the Law Enforcement Management and Administrative Statistics (LEMAS) data set will be used. It will be narrowed to California jurisdictions. We will compute the number of police officers per 1,000 residents in each jurisdiction. Clicking on the *Transform* button at the very top of the SPSS data screen and then clicking *Compute* will produce the box pictured in Figure 3.13.

In the *Target Variable* box, type the name that you want to give your new variable; in the present example, the variable will be called *policeper1000*, as the rate of officers per 1,000 residents is what the new variable will be. In the *Numeric Expression* area, type the equation you wish SPSS to follow to create the new variable. Here, the portion of the equation reading *(Police/Population)* tells the program to divide the total number of police in a jurisdiction by that jurisdiction's population size. The **1000* portion instructs SPSS to then multiply the *(Police/Population)* results by 1,000. Click *OK*, and a new variable will appear in the data screen.

The chapter review questions contain directions for accessing data sets that you can use to practice constructing charts and transforming variables into rates in SPSS. Play around with the data! Familiarize yourself with SPSS; we will be visiting it regularly throughout the book, so the more comfortable you are with it, the better prepared you will be. You do not have to worry about ruining the data set—if you change it in any way or make a mistake, just click *Don't save* when you exit the program and the file will be as good as new when you reopen it.

Figure 3.13   Transforming Variables in SPSS

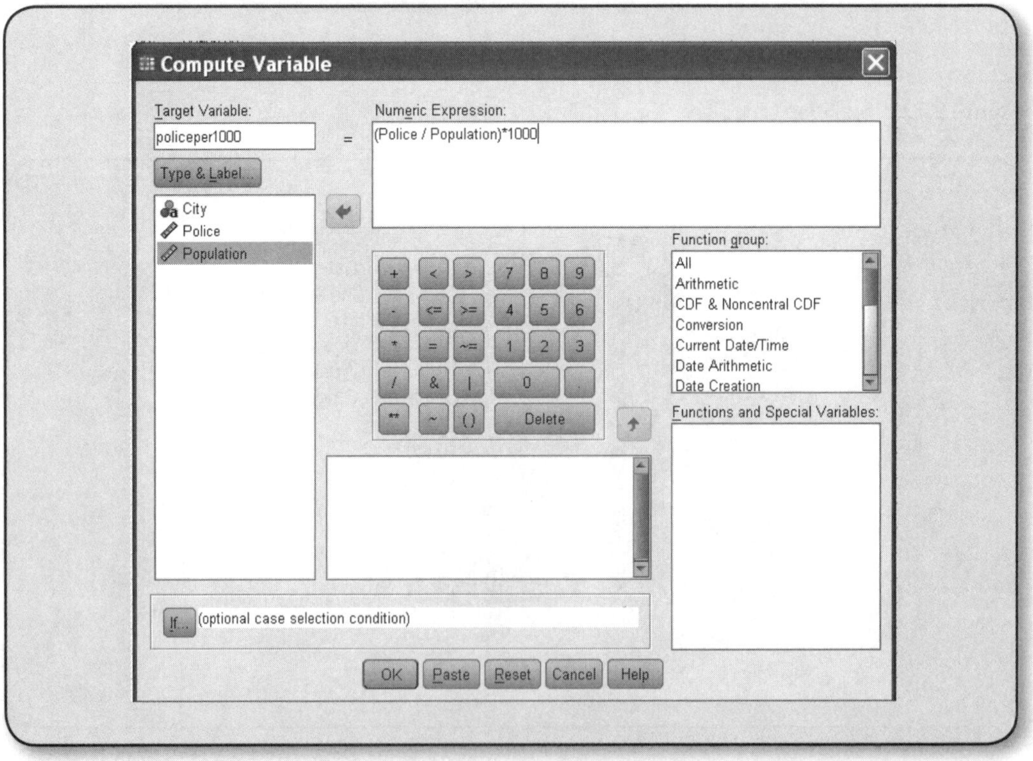

CHAPTER SUMMARY

This chapter discussed some of the most common types of graphs and charts. Frequency distributions offer basic information about the number of times certain characteristics appear in a data set. Frequencies are informative and can convey valuable information; however, numbers are often difficult to interpret whey they are in a raw format. Proportions and percentages offer a way to standardize frequencies and make it easy to determine which characteristics occur more often and which less often. Rates are another option for enhancing the interpretability of frequencies. Rates are generally multiplied by a number such as 1,000, 10,000, or 100,000.

Graphs and charts portray this same information—frequencies, proportions, percentages, and rates—using pictures rather than numbers. Pictorial representations are more engaging than their numerical counterparts and can capture audiences' interest more effectively. Pie charts may be used with categorical variables that have five or fewer classes and have been converted to percentages. Bar graphs are useful for categorical variables with any number of classes. They can be made from frequencies, proportions, percentages, or rates. For continuous data, histograms and frequency polygons can be used to graph frequencies, proportions, or percentages. Line graphs are useful for longitudinal data. Finally, some continuous variables that do not have a clear shape and are difficult to interpret in their raw form can be grouped. Grouping transforms continuous variables into ordinal ones. Histograms can be used to display data that have been grouped.

It is good to diversify a presentation by using a mix of pie charts, bar graphs, histograms, frequency polygons, and line charts. Simplicity and variety are the keys to a good presentation. Simplicity ensures that your audience can make sense of your data display quickly and easily. Variety helps keep your audience engaged and interested. Good data displays are key to summarizing data so that you and others can get a sense for what is going on in a data set.

KEY TERMS

Cell	Percentage	Classes
Univariate	Cumulative	Longitudinal variables
Frequency	Contingency table	Trends
Proportion	Bivariate	

CHAPTER 3 REVIEW PROBLEMS

1. The table below contains data from BJS' National Justice Reporting Program (NJRP). The variable is *method of conviction*, which measures whether defendants were convicted pursuant to a jury trial, a bench trial (a trial before a judge rather than a jury), or a guilty plea.

Conviction Method	*f*
Jury Trial	745
Bench Trial	686
Guilty Plea	9,837
	N = 11,268

    a. Construct columns for proportion, percentage, cumulative frequencies, cumulative proportions, and cumulative percentages.

    b. Identify the types of charts or graphs that could be used to display this variable.

    c. Based on your answer to (b), construct a graph or chart for this variable using percentages.

2. The table below contains data from the PPCS regarding female respondents' stated reasons for their most recent encounters with police.

Reason for Encounter	f
Traffic Accident	822
Traffic Stop	2,181
Reported a Crime	1,733
Requested Assistance	423
Suspect in an Investigation	97
Other	615
	N = 5,871

    a. Construct columns for proportion, percentage, cumulative frequencies, cumulative proportions, and cumulative percentages.

    b. Identify the types of charts or graphs that could be used to display this variable.

    c. Based on your answer to (b), construct a graph or chart for this variable using percentages.

3. The following table contains UCR data on the number of juveniles arrested for embezzlement for each state in 2009.

Juvenile Arrests for Embezzlement	f	Juvenile Arrests for Embezzlement	f	Juvenile Arrests for Embezzlement	f
0	3	8	1	19	2
1	7	9	1	21	1
2	2	10	3	22	1
3	4	12	3	23	1
4	3	13	1	38	1
5	2	14	1	43	1
6	5	16	2	53	1
7	3	18	1		N = 50

    a. Choose the appropriate graph type for this variable, and construct that graph using frequencies.

    b. Group this variable using 10 intervals.

    c. Choose an appropriate graph type for the grouped variable, and construct that graph using frequencies.

4. The following table contains PPCS data on the ages of respondents who said that the police had requested consent to search their car during their most recent traffic stop.

Age	f	Age	f	Age	f
16	1	29	1	46	2
17	7	30	3	47	3
18	4	31	4	49	2
19	9	32	2	50	3
20	8	34	2	51	1
21	7	35	1	52	1
22	6	37	1	53	1
23	3	38	2	54	1
24	5	39	3	55	1
25	7	41	1	57	1
26	5	42	4	58	1
27	6	44	3	59	1
28	5	45	4		N = 122

a. Choose the appropriate graph type for this variable, and construct that graph using frequencies.
b. Group this variable using 6 intervals.
c. Choose an appropriate graph type for the grouped variable, and construct that graph using frequencies.

5. The GSS contains a survey item that asks respondents whether they think that marijuana should be made legal or not. The table below contains the percentage of respondents who supported legalization in each wave of the GSS since 1990. Construct a line graph of these data, and then interpret the longitudinal trend. Does support for legalization appear to be increasing, decreasing, or staying the same over time?

Year	pct	Year	pct
1990	16.84	2000	33.54
1991	18.72	2002	35.96
1993	23.31	2004	36.41
1994	23.96	2006	36.76
1996	26.97	2008	39.78
1998	29.36		

6. The following table displays data from the official website of the U.S. courts (www.uscourts.gov) on the number of state and federal wiretap authorizations issued per year between 1997 and 2009. Construct a line graph of the data and then interpret the longitudinal trend. Have wiretap authorizations been increasing, decreasing, or staying the same over time?

Year	f	Year	f
1997	1,186	2004	1,710
1998	1,329	2005	1,773
1999	1,350	2006	1,839
2000	1,190	2007	2,208
2001	1,491	2008	1,891
2002	1,358	2009	2,376
2003	1,442		

7. The table below contains data on the number of violent crimes that occurred within seven California cities during 2009. The table also displays each city's population.

City	Violent Crimes	Population
Sacramento	4,165	470,308
Beverly Hills	81	34,506
Paso Robles	93	29,204
Bakersfield	2,099	330,897
San Francisco	5,957	809,755
Los Angeles	24,070	3,848,776
San Bernardino	1,908	199,683

   a. Compute the rate of violent crime per 1,000 city residents in each city.
   b. Select the appropriate graph type for this variable, and construct that graph using rates.

8. The table below contains data on the number of property crimes that occurred within seven California cities during 2009. The table also displays each city's population.

City	Property Crimes	Population
Sacramento	21,001	470,308
Beverly Hills	1,230	34,506
Paso Robles	927	29,204
Bakersfield	15,605	330,897
San Francisco	34,509	809,755
Los Angeles	94,240	3,848,776
San Bernardino	9,245	199,683

    a. Compute the rate of property crime per 1,000 city residents in each city.

    b. Select the appropriate graph type for this variable, and construct that graph using rates.

9. The website for this chapter (**http://www.sagepub.com/gau**) contains a data set called *Juvenile Embezzlement Arrests for Chapter 3.sav.* These are the data from Question 3 above. Use the SPSS Chart Builder to construct a frequency bar graph.

10. The file *Police in Arizona for Chapter 3.sav* contains data from LEMAS on the number of police officers per 1,000 local residents in municipal police departments in the state of Arizona. The variable *areatype* indicates whether each police agency is located in an urban area or in a rural area. Use SPSS to find the number and percentage of agencies located in each of these area types.

11. The website (**http://www.sagepub.com/gau**) also features a data file called *Hate Crimes for Chapter 3.sav.* This is the same data set used in the in-text demonstration of line graphs. Use the SPSS Chart Builder to construct a line graph mirroring the one in the text.

12. The website (**http://www.sagepub.com/gau**) contains a data set called *Conviction Type for Chapter 3.sav.* This data file contains the percentages that you computed in Question 1 above. Use the SPSS Chart Builder to construct the appropriate graph for this variable using percentages.

Chapter 4

Measures of Central Tendency

People in criminal justice and criminology are often interested in averages. Averages offer information about the centers or middles of distribution. They indicate where data points tend to cluster. This is an important thing to know. Consider the following questions that might be of interest to a researcher or practitioner in criminal justice:

1. What is the most common level of education among police officers?

2. How does the median income for people living in a structurally disadvantaged area of a certain city compare to that for all people in the city?

3. What is the average violent crime rate across all cities and towns in a particular state?

All of these questions make some reference to an average, a middle point, or, to use a more technical term, a **measure of central tendency**. Measures of central tendency offer information about where the bulk of the scores in a particular data set are located. A person who is computing a measure of central tendency is, in essence, asking, "Where is the middle?"

Measures of Central Tendency: Descriptive statistics that offer information about where the scores in a particular data set tend to cluster. Examples include the mode, median, and mean.

Measures of central tendency help speak to the issue of distribution shape, which is a concept that matters enormously in statistics. Data distributions come in many different shapes and sizes. Figure 4.1 contains data from the Law Enforcement Management and Administrative Statistics (LEMAS) survey (see Data Sources 4) on the number of sworn police officers per 100 residents in the largest jurisdictions in California. The shape this variable assumes is called a **normal distribution**. The normal curve represents an even distribution of scores. The most frequently occurring values are in the middle of the curve, and frequencies drop off as one traces the number line in the positive and negative directions.

Figure 4.1   Sworn Officers per 100 Residents in California Urban Areas (LEMAS)

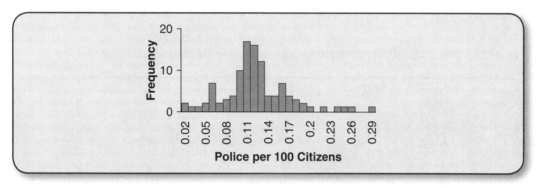

DATA SOURCES 4.1

THE LEMAS SURVEY

The Bureau of Justice Statistics conducts the Law Enforcement Management and Administrative Statistics (LEMAS) survey every 3 years in order to maintain timely descriptive data on police forces (state, county, and municipal) nationwide. Each survey cycle, BJS selects a sample of all police agencies in the U.S. and then mails them a form to fill out and return. The many variables in this data set include agency size, gender and racial breakdowns of officers in each agency, whether the agency participates in community policing, whether the agency allows its officers to carry the duty weapon of their choice, the amount of money that an agency collects per year in drug asset forfeitures, and many others.

Normal Distribution: A set of scores that clusters in the center with symmetric or roughly symmetric tails extending in each direction.

Positive Skew: A clustering of scores in the left-hand side of a distribution with some relatively large scores that pull the tail toward the positive side of the number line.

Negative Skew: A clustering of scores in the right-hand side of a distribution with some relatively small scores that pull the tail toward the negative side of the number line.

The distribution in Figure 4.2—UCR-derived homicide rates—manifests what is called a **positive skew**. Positively skewed data cluster on the left-hand side of the distribution, with extreme values in the right-hand portion that pull the tail out toward the positive side of the number line. Positively skewed data are common in criminal justice and criminology research because criminal offending and victimization is, overall, a relatively rare event. Figure 4.3 shows 2008 General Social Survey (GSS; see Data Sources 2.2) respondents' annual household incomes. This distribution has a **negative skew**: Scores cluster on the right-hand side with a tail extending toward the negative side of the number line.

Figure 4.2   Homicide Rates per 100,000 Residents in California Urban Areas (UCR)

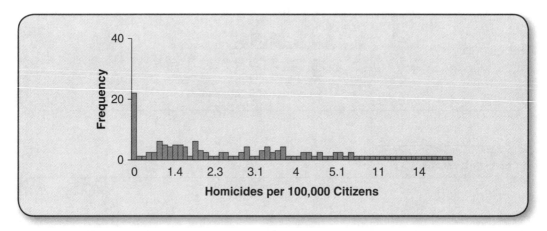

Figure 4.3   GSS Respondents' Annual Household Incomes

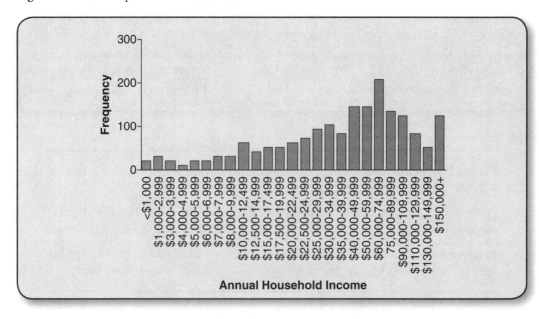

STUDY TIP: DETERMINING WHETHER SKEW IS POSITIVE OR NEGATIVE

Skew type (positive versus negative) is determined by the location of the elongated tail of a skewed distribution. Positively skewed distributions are those in which the tail extends toward the positive side of the number line; likewise, negative skew is signaled by a tail extending toward negative infinity.

Knowing whether a given distribution of data points is normal or skewed is vital in criminal justice/criminology research. Distribution shape plays a key role in inferential statistics (a.k.a. hypothesis testing) and will be something we delve into in detail in Parts II and III of the book. You must always know where the middle of your data set is located; measures of central tendency give you that information.

▣ THE MODE

The mode is the simplest of the three measures of central tendency covered in this chapter. It requires no mathematical computations and can be employed with any level of measurement; in fact, it is the only measure of central tendency available for use with nominal data. The mode is simply the most frequently occurring category or value. Table 4.1 contains data from the 2005 Police–Public Contact Survey (PPCS; Data Sources 2.1). Interviewers asked PPCS respondents whether they had been the subject of a traffic stop within the past 6 months. The 4,574 people who answered yes were then asked to report the reason for that stop. Table 4.1 presents the distribution of responses that participants gave for their stop. Before reading further, do two things as an exercise. First, identify the level of measurement of *Stop Reason*. Second, identify the reason that occurs most frequently in the data.

Table 4.1   Among Stopped PPCS Survey Respondents, Reason for the Stop

Stop Reason	*Frequency*
Speeding	2,458
Vehicle Defect	430
Record Check	479
Roadside Check for Drunk Drivers	99
Seatbelt Violation	216
Illegal Turn or Lane Change	267
Stop Sign or Light Violation	335
Other	290
Total	$N = 4,574$

If you answered *nominal* and *Speeding*, you are right! Among all survey respondents who had been stopped in the past year, the most common reason reported for a stop was a speeding violation. A percentage bar graph of the same data is shown in Figure 4.4. The mode is easily identifiable as the category accompanied by the highest bar.

Figure 4.4   Among Stopped PPCS Survey Respondents, Reason for the Stop

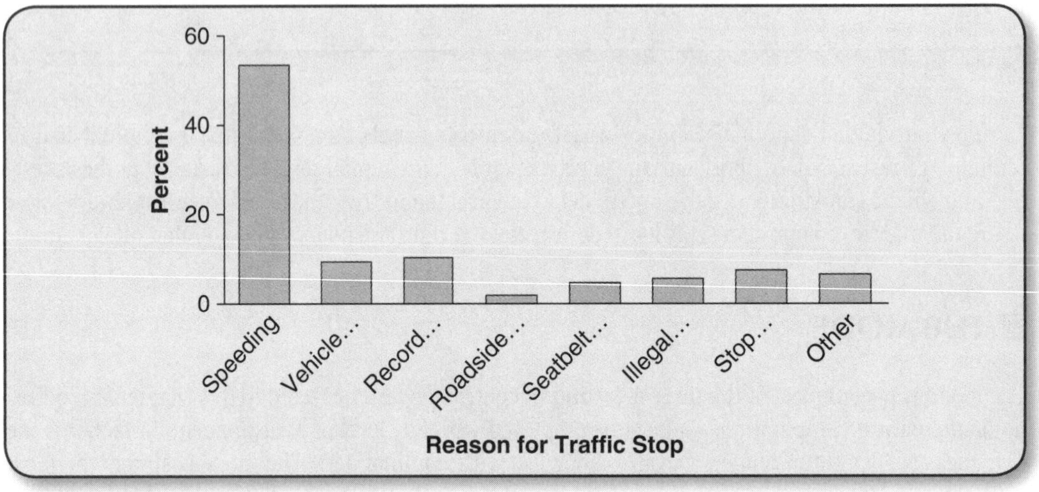

Mode: The most frequently occurring category or value in a set of scores.

RESEARCH EXAMPLE 4.1

Instant Offense and Prison Violence

Are people convicted of homicide more violent in prison than people convicted of other types of offenses? Sorensen and Cunningham (2010) analyzed the institutional conduct records of all inmates incarcerated in Florida state correctional facilities in 2003, along with the records of inmates who entered prison that same year. They divided inmates into three groups. The stock population consisted of all inmates incarcerated in Florida prisons during the year 2003, regardless of the year they were admitted into prison. The new persons admitted into prison during 2002 and serving all of 2003 comprised the admissions cohort. The close custody group was a subset of the admissions cohort and was made of the inmates who were considered to be especially high threats to institutional security. The table below contains descriptive information about the three samples. For each group, can you identify the modal custody level and conviction offense type? Visit (**http://www.sagepub.com/gau**) to view the article and see the results of this study.

Demographics	Stock Population (N = 51,527)	Admissions Cohort (N = 14,088)	Close Custody Sample (N = 4,113)
Custody level			
Community	6.0%	10.4%	0.0%
Minimum	17.2%	23.7%	0.0%
Medium	27.2%	33.8%	0.0%
Close	48.9%	32.0%	100.0%
Death row	0.6%	0.1%	0.0%
Conviction offense type			
Homicide	18.6%	5.9%	10.9%
Other violent	39.6%	33.8%	46.2%
Property	21.4%	26.7%	21.0%
Drugs	14.7%	24.0%	14.7%
Public order/weapons	5.7%	9.6%	7.2%

Source: Adapted from Table 1 in Sorensen and Cunningham (2010)

Now try finding the mode of a continuous-level variable. Table 4.2 displays Bureau of Justice Statistics (BJS; Data Sources 2.3) data on the number of death-sentenced prisoners received by state penal systems in 2008 (Snell, 2009; N sums to only 37 because in 2008, 13 states did not authorize the use of capital punishment). The left-hand column (the x column) contains all of the raw values present in the data set. The column to the right of it (the f column) shows how many times each value occurs. For instance, five states reported receiving three prisoners, one reported receiving 20, and so on. Can you identify the mode of this variable?

Table 4.2   Number of Death-Sentenced Prisoners Received by States, 2008

Number Received (x)	f
0	15
1	7
2	1
3	5
4	1
5	1
6	2
9	3
16	1
20	1
	N = 37

The mode for this data set is 0 because that value occurs 15 times, which is more often than any other score appears. Note that the mode of a categorical variable will be a category label or name, while the mode of a continuous variable will be a number. It is important to remember that the mode is the category or value that occurs the most frequently—it is *not* the frequency itself. Some variables have more than one mode if there are two or more categories that have equally high frequencies.

The above two examples highlight the relative usefulness of the mode for categorical data as opposed to continuous data. This measure of central tendency can be informative for the former, but is usually not all that interesting or useful for the latter. This is because there are other, more sophisticated measures that can be calculated with continuous data.

The strengths of the mode include the fact that this measure is simple to identify and understand. It also, as mentioned above, is the only measure of central tendency that can be used with nominal variables. The mode's major weakness is actually the flipside of its primary strength: Its simplicity means that it is usually too superficial to be of much use. It accounts for only one category or value in the data set and completely ignores the rest. It also cannot be used in more complex computations, which greatly limits its usefulness in statistics. The mode, then, can be an informative measure for nominal and possibly ordinal variables and is useful for audiences who are not schooled in statistics, but its utility is restricted, especially with continuous variables.

▣ THE MEDIAN

The **median** is another measure of central tendency. The median (*Md*) can be used with continuous and ordinal data. However, the median cannot be used with nominal data because it requires that the variable under examination be rank orderable, and nominal variables are not.

Median: The score that cuts a distribution exactly in half such that 50% of the scores are above that value and 50% are below it.

The median is the value that splits a data set exactly in half such that 50% of the data points are below it and 50% are above it. For this reason, the median is also sometimes called the *50th percentile*. The median is a positional measure, which means that it is not so much calculated as it is located. Locating the median is a three-step process. First, the categories or scores need to be rank ordered. Second, the median position (*MP*) can be computed using the formula

$$MP = \frac{(N+1)}{2}, \text{ where}$$

◁ *Formula 4(1)*

N = total sample size.

The median position tells you where the median is located within the ranked data set. The third step is to use the median position to identify the median. When *N* is odd, the median will be a value in the data set. When *N* is even, the median will have to be computed by averaging two values.

Let us compute the median homicide rate among the six urban areas in Alabama listed in Table 4.3.

Table 4.3 Homicide Rates in Six Urban Areas in Alabama

Urban Area	Homicide Rate per 100,000 Residents
Birmingham	28.59
Huntsville	7.28
Mobile	9.75
Hoover	2.78
Tuscaloosa	5.45
Montgomery	15.29
$N=6$	

The first step is to rank the rates in either ascending or descending order. Ranked in ascending order, they look like this:

$$2.78^1 \quad 5.45^2 \quad 7.28^3 \quad 9.75^4 \quad 15.29^5 \quad 28.59^6$$

Superscripts have been inserted to help emphasize the median's nature as a positional measure that is dependent upon the *location* of data points rather than these points' actual values. The superscripts represent each number's position in the data set now that the values have been rank-ordered.

Next, the formula for the median position (*MP*) will tell you where to look to find the median. Here, $MP = \dfrac{(6+1)}{2} = \dfrac{7}{2} = 3.5.$ This means that the median is in position 3.5 or, in other words, exactly halfway between positions 3 and 4. It is important to note that the MP is *not* the median; rather, it is a "map" that tells you where to look to find the median.

Finally, *MP* can be used to identify *Md*. Since the median is in position 3.5, finding its value will require averaging the numbers in positions 3 and 4 to find the exact middle between these two numbers. So the median is

$$Md = \frac{(7.28 + 9.75)}{2} = \frac{17.03}{2} = 8.52$$

Medians can also be found in ordinal-level variables, too, though the median of an ordinal variable is less precise than that of a continuous variable because the former is a category rather than a particular number. To demonstrate, we will use an item from the Police–Public Contact Survey that measures the driving frequency among respondents who said that they had been pulled over by police for a traffic stop within the past 6 months, either as a driver or as a passenger in the stopped vehicle. Table 4.4 displays the data.

The first two steps to finding the *Md* of ordinal data mirror those for continuous data. First, the categories must be rank-ordered in either ascending or descending order. You can see in Table 4.4 that the categories are already in order from the most to the least frequent driving behavior. Second, the median position must be calculated using Formula 4(1). Here, $MP = \dfrac{(5055+1)}{2} = \dfrac{5056}{2} = 2528$. The third step involves identifying the category in which the *MP* is located. You can see from the *f* column in Table 4.4

Table 4.4 Driving Frequency of PPCS Respondents Who Experienced Traffic Stops

How often do you drive?	f
Almost Every Day	4,578
A Few Days a Week	314
A Few Days a Month	51
A Few Times a Year	72
Never	40
	$N = 5,055$

that position 2,528 is located in the *Almost Every Day* group, so *Almost Every Day* is the median of this distribution. Half of respondents drive almost every day, while half drive on a less frequent basis.

The median has advantages and disadvantages. Its advantages are that it uses more information than the mode does, so it offers a more descriptive, informative picture of the data. It can be used with ordinal variables, which is advantageous because, as we will see, the mean cannot be.

A key advantage of the median is that it is not sensitive to extreme values or outliers. To understand this concept, replace Birmingham's rate of 28.59 with 16.00 in the data in Table 4.3 above and relocate the median. It did not change! That is because the median does not get pushed and pulled in various directions when there are extraordinarily high or low values in the data set. As we will see, this feature of the median gives it an edge over the mean, which *is* sensitive to extremely high or extremely low values and does shift accordingly.

The median has the disadvantage of not fully utilizing all available data points. The median offers more information than the mode does, but it still does not account for the entire array of data. This shortfall of the median can be seen by going back to the above example regarding Birmingham's homicide rate. The fact that the median did not change when the endpoint of the distribution was noticeably altered demonstrates how the median fails to offer a comprehensive picture of the entire data set. Another disadvantage of the median is that it usually cannot be employed in further statistical computations. There are limited exceptions to this rule, but generally speaking, the median cannot be plugged into statistical formulas for purposes of performing more complex analyses.

▣ THE MEAN

This brings us to the third measure of central tendency that we will cover: the **mean**. The mean is the arithmetic average of a data set. Unlike that of the median, the calculation of the mean requires the use of every raw score in a data set. Each individual point exerts a separate and independent effect on the value of the mean. The mean can be calculated only with continuous (interval or ratio) data; it cannot be used to describe categorical variables.

Mean: The arithmetic average of a set of data.

There are two formulas for the computation of the mean, each of which is for use with a particular type of data distribution. The first formula is one with which you are likely familiar from college or high school math classes. The formula is:

$$\bar{x} = \frac{\Sigma x}{N}, \text{ where}$$

Formula 4(2)

$\bar{x}$ (x bar) = the sample mean,

Σ (*sigma*) = a summation sign directing you to sum all numbers or terms to the right of it,

x = values in a given data set,

N = the sample size.

This formula tells you that to compute the mean, you must first add all the values in the data set together and then divide that sum by the total number of values. Division is required because all else being equal, larger data sets will produce larger sums, so it is vital to account for sample size when attempting to construct a composite measure such as the mean.

For the example concerning computation of the mean, we can reuse the Alabama urban homicide rate data from above (refer to Table 4.3), using the formula

$$\bar{x} = \frac{28.59 + 7.28 + 9.75 + 2.78 + 5.45 + 15.29}{6} = \frac{69.14}{6} = 11.52.$$

The second formula for the mean is used for large data sets that are organized in tabular format using both an x column that contains the raw scores, and a frequency (*f*) column that conveys information about how many times each x value occurs in the data set. Table 4.5 reproduces the death sentence data from Table 4.1 with the addition of a new column labeled *fx*. Ignore that column for a moment, and focus instead on the *x* and *f* columns.

Table 4.5 Number of Death-Sentenced Prisoners Received by States, 2008

Number Received (x)	f	fx
0	15	0
1	7	7
2	1	2
3	5	15
4	1	4
5	1	5
6	2	12
9	3	27
16	1	16
20	1	20
	$N = 37$	$\Sigma fx = 108$

Using the conventional mean formula would require extensive addition because you would have to sum a total of 37 numbers (i.e., $0 + 0 + \ldots + 1 + 1 + \ldots + 5 + 5 + \ldots + 9 + 9 + \ldots + 20$). This process is so unwieldy as to be impossible from a practical standpoint. What can be done instead is to use multiplication as a shortcut for addition; that is, instead of adding each x the number of times it occurs in the data set, the individual x values can simply be multiplied by their respective frequencies. This is what the fx column on the right-hand side of Table 4.5 is for. The mean formula for large data sets is

$$\bar{x} = \frac{\sum fx}{N}, \text{ where}$$

Formula 4(3)

f = the frequency associated with each raw score x,

fx = the product of x and f.

The difference between this formula and the one presented earlier is the addition of f, which represents the frequency. The process of computing this mean can be broken down into three steps: (1) Multiply each x by its f; (2) Sum the resulting fx products; and (3) Divide by the sample size N. Plugging the numbers from Table 4.5 into the formula, it can be seen that the mean is

$$\bar{x} = \frac{108}{37} = 2.92.$$

In 2008, each state that authorized the death penalty received an average of 2.92 death-sentenced prisoners.

STUDY TIP: HOW TO CHOOSE THE CORRECT MEAN FORMULA

Anytime you need to compute a mean, you will have to choose between Formulas 4(2) and 4(3). This is a simple enough choice if you just consider that in order to use the formula with an f in it, there must be an "f" column in the table. If there is no f column, use the formula that does not have an f.

RESEARCH EXAMPLE 4.2

Do short-term, high-rate offenders differ from long-term, low-rate offenders?

It is well known that some offenders commit a multitude of crimes over their life and others commit only a few, but the intersection of offense volume (offending rate) and time (the length of a criminal career) has received little attention from criminal justice/criminology researchers. Piquero, Sullivan, and Farrington (2010) used a longitudinal data set of males in South London who demonstrated delinquent behavior early in life and were thereafter tracked by a research team who interviewed them and looked up their official conviction records periodically until the present day. The researchers were interested in finding out whether males who committed a lot of crimes in a short amount of time (the short-term, high-rate offenders or STHR) differed significantly from those who committed crimes at a lower rate over a longer period of time (the long-term, low-rate offenders, or LTLR) on criminal justice outcomes. The researchers gathered the following descriptive statistics. The numbers not in parentheses are means. The numbers in parentheses are standard deviations, which we will learn about in the next chapter.

	Offender Type	
Variable	*LTLR (N = 44)*	*STHR (N = 21)*
Overall career length	14.5 (6.50)	10.8 (4.40)
Offenses committed per year	.42 (.31)	1.25 (.72)
Age at first conviction	17.8 (4.70)	13.5 (2.30)
Total convictions	4.9 (2.40)	11.5 (4.10)
Percentage ever incarcerated	9.1%	61.9%
Number of times incarcerated	1.46 (.50)	1.23 (.66)
Years incarcerated	1.23 (.76)	1.05 (.56)

Source: Table 1 in Piquero, Sullivan, and Farrington (2010)

You can see from the table that the long-term, low-rate offenders differed from the short-term, high-rate offenders on a number of dimensions. They were, overall, older at the time of their first arrest and had a longer mean career length. They committed many fewer crimes per year and were much less likely to have been sentenced to prison.

Piquero et al.'s analysis reveals a dilemma about what should be done about these groups of offenders with respect to sentencing; namely, it shows how complicated the question of imprisonment is. The short-term, high-rate offenders might seem to be the best candidates for incarceration based on their volume of criminal activity, but these offenders' criminal careers are quite short. It makes no sense from a policy and budgetary perspective to imprison a bunch of people who would not be committing crimes if they were free in society. The STHR offenders also tended to commit property offenses rather than violent ones. The long-term, low-rate offenders, by contrast, committed a disproportionate number of violent offenses despite the fact that their overall number of lifetime offenses was lower than that for the STHR group. Again, though, the question of the utility of imprisonment arises: Is it worth incarcerating someone who, though he may still have many years left in his criminal career, will commit very few crimes during that career? The dilemma of sentencing involves the balance between public safety and the need to be very careful in the allotting of scarce correctional resources.

The mean is sensitive to extreme values and outliers, which gives it both an advantage and a disadvantage relative to the median. The advantage is that the mean uses the entire data set and accounts even for the "weird" values that may appear at the high or low ends of the distribution. The median, by contrast, ignores these values. The disadvantage is that the mean's sensitivity to extreme values makes this measure somewhat unstable; it is vulnerable to the disproportionate impact that a small number of extreme values can exert on the data set. The following example illustrates this property of the mean.

Trace the changes in the mean homicide rates from left to right in Table 4.7. Do you notice how the rate increases with the successive introductions of San Bernardino and Oakland? This is because these two cities have noticeably higher homicide rates than the other cities in the table, and the introduction of them into the table inflates the mean.

Table 4.7   UCR-Based Homicide Rates in Urban Areas of California

Urban Area	Homicide Rate per 100,000 Residents	Homicide Rate per 100,000 Residents	Homicide Rate per 100,000 Residents
San Diego	3.12	3.12	3.12
Fresno	8.73	8.73	8.73
Los Angeles	8.11	8.11	8.11
San Bernardino	—	16.03	16.03
Oakland	—	—	25.71
	$N=3$	$N=4$	$N=5$
	$\bar{x}=6.65$	$\bar{x}=9.00$	$\bar{x}=12.34$

STUDY TIP: CHOOSING A MEASURE OF CENTRAL TENDENCY

The primary criterion on which you must rely when deciding what measure of central tendency to use is a given variable's level of measurement. The following chart offers a summary.

	Mode	Median	Mean
Nominal	✓		
Ordinal	✓	✓	
Grouped	✓	✓	✓
Interval	✓	✓	✓
Ratio	✓	✓	✓

▣ USING THE MEAN AND MEDIAN TO DETERMINE DISTRIBUTION SHAPE

Given that the median is invulnerable to extreme values but the mean is not, the best strategy is to report both of these measures when describing data shape. The mean and median can, in fact, be compared to form a judgment about whether the data are normally distributed, positively skewed, or negatively skewed. In normal distributions, the mean and median will be approximately equal. Positively skewed distributions will have means markedly greater than their medians. This is because extremely high values in positively skewed distributions pull the mean up but do not affect the location of the median. Negatively skewed distributions, on the other hand, will have medians that are noticeably larger than their means because extremely low numbers tug the mean downward but do not alter the median's value. Figure 4.5 illustrates this conceptually.

Figure 4.5 The Mean and Median as Indicators of Distribution Shape

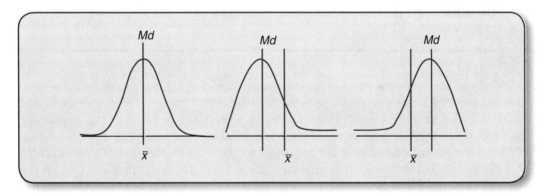

To give a full picture of a data distribution, then, it is best to make a habit of reporting both the mean and the median.

The mean—unlike the mode or median—forms the basis for further computations; in fact, the mean is an analytical staple of many inferential hypothesis tests. The reason that the mean can be used in this manner is that the mean is the **midpoint of the magnitudes**. This point merits its own section.

Midpoint of the Magnitudes: The property of the mean that causes all deviation scores based on the mean to sum to zero.

▣ THE MEAN AS THE MIDPOINT OF THE MAGNITUDES

The mean possesses a vital property that enables its use in complex statistical formulas. To understand this, we must first discuss **deviation scores**. A deviation score is a given data point's distance from its group mean. The formula for a deviation score is based on simple subtraction:

$$d_i = x_i - \bar{x}, \text{ where}$$

> Formula 4(4)

x_i = a given data point,

d_i = the deviation score for a given data point x_i,

$\bar{x}$ = the sample mean.

Suppose, for instance, that a group's mean is $\bar{x} = 24$. If a certain raw score x_i is 22, then $d_{x=22} = 22 - 24 = -2$. A raw score of 25, by contrast, would have a deviation score of $d_{x=25} = 25 - 24 = 1$.

Deviation Score: The distance between the mean of a data set and any given raw score in that set.

A deviation score embodies two pieces of information. The first is the absolute value of the score or, in other words, how far from the mean a particular raw score is. Data points that are exactly equal to the mean will have deviation scores of 0; therefore, deviation scores with larger absolute values are farther away from the mean, while deviation scores with smaller absolute values are closer to it.

The second piece of information that a deviation score conveys is whether the raw score associated with that deviation score is greater than or less than the mean. Positive deviation scores represent raw scores that are greater than the mean and negative deviation scores signify raw numbers that are less than the mean. You can thus discern two things about the raw score x_i that a given deviation score d_i represents: (1) the distance between x_i and $\bar{x}$; and (2) whether x_i is above $\bar{x}$ or below it. Notice that you would not even need to know the actual value of x_i or $\bar{x}$ in order to effectively interpret a deviation score. Deviation scores convey information about the position of a given data point with respect to its group mean; that is, deviation scores offer information about raw scores' *relative*, rather than absolute, positions within their group. Figure 4.6 illustrates this.

What lends the mean its title as the midpoint of the magnitudes is the fact that deviation scores computed using the mean (as opposed to the mode or median) always sum to zero. As Salkind (2008) put it, "The mean is like the fulcrum on a seesaw. It's the centermost point where all the values on one side of the mean are equal in weight to all the values on the other side of the mean" (p. 21). The seesaw analogy is apt—the mean is the value in the data set at which all values below it balance out with all values above it. For an example of this, try summing the deviation scores in Figure 4.6 below. What is the result?

To demonstrate this concept more concretely, homicide data from Table 4.7 are reproduced in Table 4.8. The d column contains the deviation score for each of the raw homicide rates.

Figure 4.6 Deviation Scores in a Set of Data With a Mean of 24

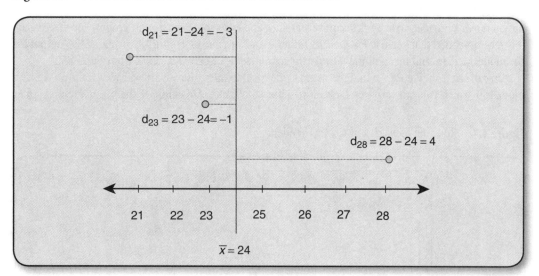

Table 4.8 UCR-Based Homicide Rates in Urban Areas of California

Urban Area	Homicide Rate	d
San Diego	4.33	4.33 – 13.46 = –9.13
Fresno	8.41	8.41 – 13.46 = –5.05
Los Angeles	9.97	9.97 – 13.46 = –3.49
San Bernardino	15.95	15.95 – 13.46 = 2.49
Oakland	28.64	28.64 – 13.46 = 15.18
	$\bar{x} = 13.46$	$\Sigma = .00$

Illustrative of the mean as the fulcrum of the data set, the positive and negative deviation scores ultimately cancel each other out, as can be seen by the sum of zero at the bottom of the deviation score column. This represents the mean's property of being the midpoint of the magnitudes—it is the value that perfectly balances all of the raw scores. This characteristic is what makes the mean a central component to most statistical analyses.

▣ SPSS

Criminal justice and criminology researchers generally work with large data sets, so computing measures of central tendency by hand is not feasible; luckily, it is not necessary, either, because statistical programs such as SPSS can be used instead. There are two different ways to obtain central tendency output.

Under the *Analyze* → *Descriptive Statistics* menu, SPSS offers the options *Descriptives* and *Frequencies* (see Figure 4.7). Both of these functions will produce central tendency analyses, but the *Frequencies* option offers a broader array of descriptive statistics and even some charts and graphs. For this reason, it is recommended that you use *Frequencies* rather than *Descriptives*. Figure 4.8 displays SPSS output for the mode, median, and mean of the *death-sentenced prisoners received* variable from Table 4.5.

You can see from Figure 4.8 that the mean is identical to the result we arrived at by hand above. The mode is 0, which you can verify by looking at Table 4.5. We can also compare the mean and median to

Figure 4.7   Running Measures of Central Tendency in SPSS

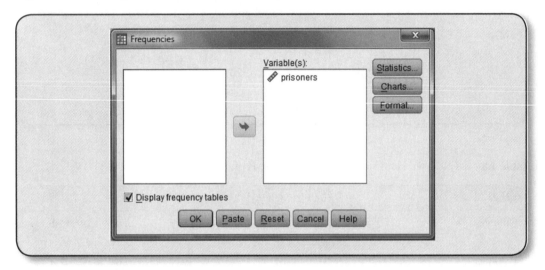

Figure 4.8   SPSS Frequencies Output

Statistics

Number of death-sentenced prisoners received, 2008

N	Valid	37
	Missing	0
Mean		2.9189
Median		1.0000
Mode		.00

determine the shape of the distribution. With a mean of 2.92 and a median of 1.00, do you think that this distribution is normally distributed, positively skewed, or negatively skewed? If you said positively skewed, you are correct!

CHAPTER SUMMARY

This chapter introduced you to three measures of central tendency: the mode, median, and mean. These statistics offer summary information about the middle or average score in a data set. The mode is the most frequently occurring category or value in a data set. The mode can be used with variables of any measurement type (nominal, ordinal, interval, or ratio) and is the only measure that can be used with nominal variables. Its main weakness is in its simplicity and superficiality—it is generally not all that useful.

The median is a better measure than the mode for data measured at the ordinal or continuous level. The median is the value that splits a data set exactly in half. Since it is a positional measure, the median's value is not affected by the presence of extreme values—this makes the median a better reflection of the center of a distribution the mean is when a distribution is highly skewed. The median, though, does not utilize all data points in a distribution, which makes it less informative than the mean.

The mean is the arithmetic average of the data and is for use with continuous variables only. The mean accounts for all values in a data set, which is good because no data are omitted; however, the flipside is that the mean is susceptible to being pushed and pulled by extreme values.

It is good to report both the mean and the median because they can be compared to determine the shape of a distribution. In a normal distribution, these two statistics will be approximately equal. In a positively skewed distribution, the mean will be markedly greater than the median, and in a negatively skewed distribution, the mean will be noticeably smaller than the median. Reporting both of them provides your audience with much more information than they would have if you just reported one or the other.

The mode, median, and mean can all be obtained in SPSS using the *Analyze* → *Descriptive Statistics* → *Frequencies* sequence. As always, GIGO! When you order SPSS to produce a measure of central tendency, it is your responsibility to ensure that the measure you choose is appropriate to the variable's level of measurement. If you err, SPSS will probably not alert you to the mistake—you will get output that looks fine but is actually garbage. Be careful!

CHAPTER 4 REVIEW PROBLEMS

1. A survey item asks respondents "How many times have you shoplifted?" and allows them to fill in the appropriate number.
   a. What level of measurement is this variable?
   b. What measure or measures of central tendency can be computed on this variable?

2. A survey item asks respondents "How many times have you shoplifted?" and gives them the answer options: *0; 1 – 3; 4 – 6; 7 or more.*
   a. What level of measurement is this variable?
   b. What measure or measures of central tendency can be computed on this variable?

3. A survey item asks respondents "Have you ever shoplifted?" and tells them to circle *yes* or *no*.

   a. What level of measurement is this variable?
   b. What measure or measures of central tendency can be computed on this variable?

4. Explain what an extreme value is. Include in your answer (1) the effect extreme values have on the median, if any; and (2) the effect extreme values have on the mean, if any.

5. Explain why the mean is the midpoint of the magnitudes. Include in your answer (1) what deviation scores are and how they are calculated, and (2) what deviation scores always sum to.

6. In a negatively skewed distribution

   a. the mean is less than the median.
   b. the mean is greater than the median.
   c. the mean and median are approximately equal.

7. In a normal distribution

   a. the mean is less than the median.
   b. the mean is greater than the median.
   c. the mean and median are approximately equal.

8. In a positively skewed distribution

   a. the mean is less than the median.
   b. the mean is greater than the median.
   c. the mean and median are approximately equal.

9. In a positively skewed distribution, the tail extends toward _____ of the number line.

   a. the positive side
   b. both sides
   c. the negative side

10. In a positively skewed distribution, the tail extends toward _____ of the number line.

    a. the positive side
    b. both sides
    c. the negative side

11. The table below contains UCR data on the relationship between murder victims and their killers.

Victim-Offender Relationships Among Murder Victims in 2008 (UCR)

Offender's Relationship to Victim	f
Family Member	1,145
Intimate Partner	1,333
Friend	504
Acquaintance	3,068
Other Nonstranger	120
Stranger	1,742
	$N = 7,912$

a. Identify this variable's level of measurement and, based on that, state the appropriate measure or measures of central tendency.
b. Determine or calculate the measure or measures of central tendency that you identified in part (a).

12. The frequency distribution below contains violent victimization data from the Bureau of Justice Statistics' report based on NCVS data (Harrell, 2007). Use this table to do the following.

NCVS-Based Annual Violent Victimization Rate per 1,000 Persons

Race	Victimization Rate
Black/African American	28.70
White	22.80
American Indian/Alaska Native	56.80
Asian/Pacific Islander	10.60
Hispanic/Latino	24.30
$N = 5$	

a. Identify the median victimization rate using all three steps.
b. Compute the mean victimization rate across all racial groups. Remember that there are two formulas for the mean; the first thing you need to do is decide which one is appropriate for this data distribution.

13. The following frequency distribution below shows data on the number of female American Indian officers on police forces nationwide.

Female American Indian Officers in the LEMAS Survey

Number	f	Number	f
0	2,829	8	1
1	112	9	1
2	14	11	1
3	12	12	1
4	7	60	1
5	4		$N = 2,985$
7	2		

a. Identify the mode.
b. Compute the mean. Remember that there are two formulas for the mean; the first thing you need to do is decide which one is appropriate for this data distribution.

UCR-Based Robbery Rates in Urban Areas of Florida

Urban Area	Robbery Rate
Ft. Lauderdale	37.44
Jacksonville	29.12
Gainesville	20.74
Orlando	32.62
Pensacola	18.48
Miami	49.95
N = 6	

14. Use the 2009 Florida UCR data in the table below to do the following.
   a. Identify the median robbery rate using all three steps.
   b. Calculate the mean robbery rate.
   c. Compare the mean and median to determine the shape of this distribution. Is it normal or skewed and, if skewed, in what direction is the skew? How can you tell?

15. Morgan, Morgan, and Boba (2010) report state and local government expenditures, by state, for police protection in the year 2007. The data below contain a random sample of states and the dollars spent per capita in each state for police services.

Dollars Spent Per Capita on Police Protection, 2007

State	Dollars
Illinois	317
Arkansas	170
Alabama	211
Ohio	258
Washington	219
California	381
Florida	345
Maine	176
New Jersey	353
Texas	220
N = 10	

a. Identify the median dollar amount using all three steps.
b. Calculate the mean dollar amount.

   c. Compare the mean and median to determine the shape of this distribution. Is it normal or skewed and, if skewed, in what direction is the skew? How can you tell?

16. Morgan et al. (2010) also report violent crime arrest rates for juveniles in 2008 (measured as the number of juveniles arrested for violent crimes per 100,000 juveniles).

Juvenile Violent Crime Arrest Rate, 2008

State	Rate
Nevada	3.90
Louisiana	5.40
California	5.40
Nebraska	2.80
Oklahoma	4.90
Arizona	4.10
New Hampshire	0.00
Maryland	5.10
$N = 8$	

   a. Identify the median arrest rate using all three steps.
   b. Calculate the mean arrest rate.
   c. Compare the mean and median to determine the shape of this distribution. Is it normal or skewed and, if skewed, in what direction is the skew? How can you tell?

17. The data set *Weapons on School Property for Chapter 4.sav* (**http://www.sagepub.com/gau**) contains high school students' reports about whether they carried a weapon on school grounds in 2007 (Morgan et al., 2010).

   a. Run an SPSS analysis to determine the median and mean of this variable.
   b. Compare the mean and median to determine the shape of this distribution. Is it normal or skewed and, if skewed, in what direction is the skew? How can you tell?

18. The data set *Handgun Murder Rates for Chapter 4.sav* (**http://www.sagepub.com/gau**) contains the 2009 handgun murder rates by state, as reported by the Uniform Crime Statistics.

   a. Run an SPSS analysis to determine the median and mean of this variable.
   b. Compare the mean and median to determine the shape of this distribution. Is it normal or skewed and, if skewed, in what direction is the skew? How can you tell?

KEY TERMS

Measures of central tendency	Median	Midpoint of the magnitudes
Mode	Mean	Deviation score

GLOSSARY OF SYMBOLS INTRODUCED IN THIS CHAPTER

MP	Median position
Md	Median
$\bar{x}$	Mean
d_i	Deviation score for a raw score x_i

REFERENCES

Harrell, E. (2007). *Black victims of violent crime* (BJS Publication No. NCJ 214258). Washington, DC: U.S. Department of Justice.

Morgan, K. O., Morgan, S., & Boba, R. (2010). *Crime: State rankings 2010.* Washington, DC: CQ Press.

Piquero, A. R., Sullivan, C. J., & Farrington, D. P. (2010). Assessing differences between short-term, high-rate offenders and long-term, low-rate offenders. *Criminal Justice and Behavior, 37*(12), 1309–1329.

Salkind, N.J. (2008). *Statistics for people who (think they) hate statistics* (3rd ed.). Los Angeles: Sage.

Snell, T.L. (2009). *Capital punishment, 2008—Statistical tables* (BJS Publication No. NCJ 228662). Washington, DC: U.S. Department of Justice.

Sorensen, J., & Cunningham, M.D. (2010). Conviction offense and prison violence: A comparative study of murderers and other offenders. *Crime & Delinquency, 56*(1), 103–125.

Chapter 5

Measures of Dispersion

C onsider this question: Do you think that there is more variability in the physical size of house cats or in that of dogs? In other words, if I gathered a random sample of house cats and a random sample of dogs, would I find a greater diversity of sizes and weights in the cat sample or in the dog sample? In the dog sample, of course! Dogs range from puny things like miniature Chihuahuas all the way up to hulking creatures like Great Danes. Cats, on the other hand, are ... well, cat-sized. If I were to draw separate "physical size" distributions for dogs and house cats, they might look something like Figure 5.1.

Figure 5.1 Hypothetical Distributions of Dog and House Cat Sizes

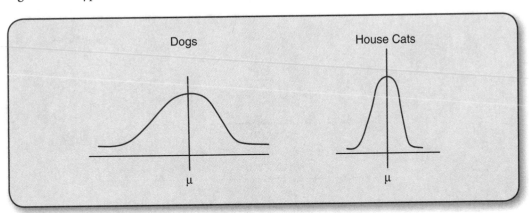

The dog distribution would be wider and flatter than the cat distribution, while the cat distribution would appear somewhat tall and narrow as compared to the dog distribution.

Now, let us add a twist and assume that the two distributions in question have the same mean. (The dog and house cat distributions would, of course, have different means.) Consider the hypothetical raw data for variables X_1 and X_2 in Table 5.1.

Table 5.1   Hypothetical Data in Two Samples

Sample 1	Sample 2
10	0
11	15
9	20
10	5
10	10
$\bar{x}_1 = 10$	$\bar{x}_2 = 10$

These distributions have the same mean, so if all you knew about each of them was their mean, you might be tempted to conclude that they are similar distributions. This conclusion would be quite wrong, though. Look at Figure 5.2, which displays a line chart for these two variables. Which series do you think represents Sample 1? Sample 2? If you said that the stars are Sample 1 and diamonds are Sample 2, you are correct. You can see that the raw scores of Sample 1 cluster quite tightly around the mean, while Sample 2's scores are scattered about in a much less cohesive manner.

Figure 5.2   Line Chart for Hypothetical Variables X_1 and X_2

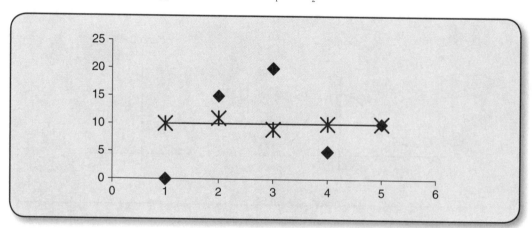

Dispersion: The amount of spread or variability among the scores in a distribution.

The above examples highlight the importance of the subject of this chapter: measures of dispersion. **Dispersion** (sometimes also called *variability*) is the amount of "spread" present in a set of raw scores. Dogs have more dispersion in physical size than house cats have, just as Sample 2 has more dispersion than Sample 1 has. Measures of dispersion are vital from an informational standpoint for the reason exemplified by the thought experiment just presented—measures of central tendency convey only so much information about a distribution and can actually be misleading, because it is possible for two very discrepant distributions to have similar means. Measures of dispersion thus go hand in hand with measures of central tendency, and the two types of descriptive statistics are usually presented alongside one another. This point will be revisited at the end of the chapter after four of the most common measures of central tendency are discussed: the variation ratio, range, variance, and standard deviation.

▣ THE VARIATION RATIO

The **variation ratio** (*VR*) is the only measure of dispersion discussed in this chapter that can be used with nominal and ordinal data; the remainder of the measures covered here can only be employed when a variable is measured at either the interval or ratio level. The *VR* is based on the mode and measures the proportion of cases that are *not* in the modal category. Recall from the discussion of proportions in Chapter 3 that the proportion of cases that are in a certain category can be found using

the formula $p = \frac{f}{N}$. It is easy to take this formula a step further in order to calculate the proportion that is not in a particular category.

Variation ratio: A measure of dispersion for variables of any level of measurement that is calculated as the proportion of cases located outside the modal category.

Bounding rule: The rule stating that all proportions range from 0.00 to 1.00.

Rule of the complement: Based on the bounding rule, the rule stating that the proportion of cases that are not in a certain category can be found by subtracting the proportion that are in that category from 1.00.

For this inquiry, we rely on the **bounding rule**, which states that proportions always range from 0.00 to 1.00. The bounding rule leads to the **rule of the complement**. This rule states that the proportion of cases in a certain category (call this category *A*) and the proportion located in other categories (call this *Not A*) always sum to 1.00. Formally:

$$p(A) + p(Not\ A) = 1.00.$$

If $p(A)$ is known, the formula can be reconfigured thusly in order to calculate $p(Not\ A)$:

$$p(Not\ A) = 1.00 - p(A).$$

Formula 5(1)

STUDY TIP: REMEMBER THE BOUNDING RULE AND RULE OF THE COMPLEMENT

Pay close attention to these concepts! They will resurface in Chapter 6.

The variation ratio is a spinoff of the rule of the complement:

$$VR = 1.00 - \frac{f_{mode}}{N}, \text{ where}$$

Formula 5(2)

f_{mode} = the number of cases in the modal category; and

N = the sample size.

To illustrate use of the VR, consider Table 5.2, which contains data from the National Judicial Reporting Program (NJRP; see Data Sources 3.1 in Chapter 3) on the method of case disposition for people convicted of drug trafficking. (Note: Defendants with unknown disposition modalities were omitted.)

Table 5.2 Case Disposition Method for Persons Convicted of Drug Trafficking

Method	f
Jury Trial	746
Bench Trial	646
Guilty Plea	53,020
	$N = 54,412$

To compute the VR, you must first identify the mode and its associated frequency. Here, the mode is *Guilty Plea* and its frequency is 53,020. Now, plug the numbers into Formula 5(2):

$$VR = 1.00 - \frac{53020}{54412} = 1.00 - .97 = .03.$$

The variation ratio is .03, which is a very small proportion—virtually zero—and indicates that the vast majority of the cases are in the modal category, with relatively few outside of it.

Table 5.3 Probation Sentences Received Among Persons
Convicted of Drug Trafficking

Probation?	f
Yes	43,375
No	50,647
	N = 94,022

Contrast that to Table 5.3, also displaying NJRP data. The variable in this table measures whether or not people convicted of drug trafficking received a probation sentence (alone or in combination with a prison or jail term).

The mode is *No* and the variation ratio is

$$VR = 1.00 - \frac{50647}{94022} = 1.00 - .54 = .46.$$

It is easy to see that this *VR* is much larger than that for the *disposition method* variable; .46 is quite close to .50, indicating that the sample was split nearly in half with respect to whether or not a probation sentence was imposed.

In sum, the variation ratio offers information about whether the data tend to cluster inside the modal category or a fair number of the cases are in other categories. This statistic can be used with nominal and ordinal variables, which makes it unique relative to the range, variance, and standard deviation. The problem with the *VR* is that it uses only a small portion of the data available in a distribution. It indicates the proportion of cases not in the modal category, but it does not actually show where those cases *are* located. This makes interpretation of the *VR* tricky when variables have three or more classes. It would be nice to know where the data are rather than merely where they are not.

Range: A measure of dispersion for continuous variables that is calculated by subtracting the smallest score from the largest.

▣ THE RANGE

The **range** (*R*) is the simplest measure of dispersion for continuous-level variables. It measures the span of the data or, in other words, the distance between the smallest and largest values. The range is very easy to compute:

$$R = x_{maximum} - x_{minimum}.$$

 Formula 5(3)

The first step in computing the range is identification of the maximum and minimum values in the data set. The minimum is then subtracted from the maximum, as shown in Formula 5(3). That is all there is to it!

For an example, we can use data from the Uniform Crime Reports (UCR; Data Sources 1.1) on the number of juveniles arrested on suspicion of homicide in 2009 in northeastern states. Table 5.4 shows the data.

Table 5.4   Number of Juveniles Arrested for Homicide by Northeastern State, 2009

State	Arrests
Maine	1
New Hampshire	0
Vermont	0
Massachusetts	4
Rhode Island	0
Connecticut	7
New York	32
Pennsylvania	36
New Jersey	28
$N = 9$	

To calculate the range, first identify the maximum and minimum values. The maximum is 36 and the minimum is 0. The range, then, is

$$R = 36 - 0 = 36.$$

The range is also easy to find with large data sets organized into frequency distributions. Table 5.5 contains data from the 2005 Police–Public Contact Survey (PPCS; see Data Sources 2.1). The table displays the number of face-to-face contacts with police in the past 6 months reported by those respondents who said that they had had at least one contact. Finding the range in a frequency distribution like this one may seem more complicated, but it is actually no different than computing it for a smaller distribution like that in Table 5.3. The highest score is 96 and the smallest is 1, so the range is

$$R = 96 - 1 = 95.$$

It is important to remember that the range is computed using the actual raw scores, not the frequencies of those scores. Always make sure that you are looking at the correct column when calculating the range.

Table 5.5 Number of Face-to-Face Contacts With Police Reported by PPCS Respondents Who Had Experienced Contact

Number	f	Number	f	Number	f
1	8,577	10	41	25	4
2	2,027	11	3	26	1
3	626	12	27	30	2
4	246	13	2	40	1
5	124	15	8	50	3
6	76	17	1	60	1
7	22	20	13	96	3
8	23	21	2	N = 11,843	
9	8	24	2		

The range has advantages and disadvantages. First, it is a nice measure because it is so simple and straightforward. It offers useful information and is extremely easy to calculate and understand. Its strength, though, is also its major weakness—it is too simplistic to be of much use. The range is a very superficial measure of dispersion and offers only the most minimal information about a variable. Just by looking at the range for the PPCS data, for instance, all you know is that there are 95 units between the largest and smallest values in the data set. If you did not have access to the raw data in Table 5.5, you would have absolutely no idea what the maximum and minimum numbers are. Even more troubling, you would have no information about their distribution. Are they normally distributed? Are they skewed? Do they cluster close to the mean, or are they very spread out? There is no way to access any of this vital information merely by knowing the range of a data set. Also, because the range does not use all of the available data, it has no place in further computations.

◙ THE VARIANCE

Like the range, the **variance** can only be used with continuous data; unlike the range, the variance utilizes every number in a data set. The variance and its offshoot, the standard deviation (discussed next), are the quintessential measures of dispersion. You will see them repeatedly throughout the remainder of the book. For this reason, it is crucial that you develop a comprehensive understanding of them both formulaically and conceptually.

The formula for the variance can appear rather intimidating, so let us work our way up to it piece by piece and gradually construct it. We will use the *Number of Juveniles Arrested for Homicide* data from Table 5.4 to illustrate the computation as we go. First, you should recall the concept of deviation scores that was discussed in the last chapter. Go back and review if necessary.

A deviation score is the difference between a raw score in a distribution and that distribution's mean, as such:

$$d_i = x_i - \bar{x}, \text{ where}$$

x_i = a given data point,

d_i = the deviation score for a given data point x_i,

$\bar{x}$ = the sample mean.

The variance is constructed from mean-based deviation scores. The first step in computing the variance is to compute the mean, and the second is to find the deviation score for each raw value in a data set. The first piece of the variance formula, therefore, is $x_i - \bar{x}$ for every raw score x_i in the data. Table 5.6a shows the original data from Table 5.5 along with a deviation score column to the right of the raw scores. The mean of the data set has been calculated. In the deviation score column, the mean has been subtracted from each raw score to produce a variety of positive and negative deviation scores.

Table 5.6a   Number of Juveniles Arrested for Homicide in Northeastern States, 2009

State	Arrests	$x_i - \bar{x}$
Maine	1	$1 - 12 = -11$
New Hampshire	0	$0 - 12 = -12$
Vermont	0	$0 - 12 = -12$
Massachusetts	4	$4 - 12 = -8$
Rhode Island	0	$0 - 12 = -12$
Connecticut	7	$7 - 12 = -5$
New York	32	$32 - 12 = 20$
Pennsylvania	36	$36 - 12 = 24$
New Jersey	28	$28 - 12 = 16$
	$\bar{x} = 12$	

Deviation scores are a good first step, but what we end up with is an array of numbers. A table full of deviation scores is no more informative than a table full of raw scores. What we need is a summary statistic of some kind, a single number that represents all of the individual deviation scores. The most obvious measure is the sum—sums are good ways of packaging multiple numbers into a single numerical term. The problem with this approach, though, should be obvious: As discussed in Chapter 4, deviation scores based on the mean of a data set sum to zero. The sum of the $x_i - \bar{x}$ column will, therefore, always be zero or within rounding error of it, which makes summing useless as a measure of variance.

STUDY TIP: CHECK YOUR MATH!

Since the sum of the deviation scores should always be zero or within rounding error of it, you can check your math up to this stage of the variance computation procedure by ensuring that the sum of your calculated deviation scores is either exactly or approximately zero.

We need to find a way to get rid of those pesky negative signs. If all of the numbers were positive, it would be impossible for the sum to be zero. Squaring is used to accomplish this objective. Squaring each deviation score eliminates the negative signs (because negative numbers always become positive when squared), and, as long as the squaring is applied to all of the scores, it is not a problematic transformation of the numbers. Table 5.6b contains a new right-hand column showing the squared version of each deviation score.

Table 5.6b   Number of Juveniles Arrested for Homicide in Northeastern States, 2009

State	*Arrests*	$x_i - \bar{x}$	$(x_i - \bar{x})^2$
Maine	1	$1 - 12 = -11$	$(-11)^2 = 121$
New Hampshire	0	$0 - 12 = -12$	$(-12)^2 = 144$
Vermont	0	$0 - 12 = -12$	$(-12)^2 = 144$
Massachusetts	4	$4 - 12 = -8$	$(-8)^2 = 64$
Rhode Island	0	$0 - 12 = -12$	$(-12)^2 = 144$
Connecticut	7	$7 - 12 = -5$	$(-5)^2 = 25$
New York	32	$32 - 12 = 20$	$20^2 = 400$
Pennsylvania	36	$36 - 12 = 24$	$24^2 = 576$
New Jersey	28	$28 - 12 = 16$	$16^2 = 256$
	$\bar{x} = 12.00$	$\Sigma = 0.00$	

Now the numbers can be summed. Recall that the symbol Σ (*sigma*) tells you to sum whatever is to the right of it. We can thus write the sum of the right-hand column in Table 5.6b as $\Sigma(x_i - \bar{x})^2$ and we can compute the answer as such:

$$\Sigma(x_i - \bar{x})^2 = 121 + 144 + 144 + 64 + 144 + 25 + 400 + 576 + 256 = 1,874.$$

This sum represents the total squared deviations from the mean. We are not quite done yet, though, because sample size must be taken into consideration in order to control for the number of scores present in the variance computation. The sum of the squared deviation scores must be divided by the sample size.

There is a bit of a hiccup, however: The variance formula for samples tends to produce an estimate of the population variance that is downwardly biased (i.e., too small) because samples are littler than populations and therefore contain less variability. This problem is especially evident in very small samples, such as when $N < 50$.

The way to correct for this bias in sample-based estimates of population variances is to subtract 1.00 from the sample size in the formula for the variance. The sample variance is symbolized s^2, whereas the population variance is symbolized σ^2, which is a lowercase sigma. The reason for the exponent is that, as described above, the variance is composed of squared deviation scores. The symbol s^2 thus signifies the squared nature of this statistic.

STUDY TIP: VARIANCE IS ALWAYS POSITIVE

The exponent in the symbol for the variance (s^2) should remind you that the variance is a squared number and that since squared numbers can never be negative, the variance will always be positive. If you compute the variance on a practice problem, homework assignment, or exam and obtain a final answer that is negative, then you need to go back and check your math because there is an error in your calculations.

We can now assemble the entire s^2 formula and compute the variance of the juvenile homicide arrest data. The formula for the sample variance is

$$s^2 = \frac{\sum (x_i - \bar{x})^2}{N-1}.$$

Formula 5(4)

Plugging in the numbers and solving yields

$$s^2 = \frac{1874}{9-1} = 234.25.$$

The variance of the juvenile homicide arrest data is 234.25. This is the average squared deviation from the mean in this data set.

The variance is preferable to the variation ratio and range because it uses all of the raw scores in a data set and is therefore a more informative measure of dispersion. The variance is thus a more informative measure. When the data are continuous, the variance is better than either the *VR* or the range.

▣ THE STANDARD DEVIATION

Despite its usefulness, the variance has an unfortunate hitch. When we squared the deviations (refer to Table 5.6b), we by definition also squared the *units* in which those deviations were measured. The variance of the juvenile homicide arrest data, then, is 234.25 *arrests squared*. This obviously makes no sense. The variance produces oddities such as *crimes squared* and *years squared*. Something needs to be done to correct this. Luckily, a solution is at hand. Since the problem was created by squaring, it can be solved by doing the opposite of squaring—taking the square root. The square root of the variance (symbolized *s*) is the **standard deviation**:

$$s = \sqrt{\frac{\sum \left(x_i - \overline{x}\right)^2}{N-1}} = \sqrt{s^2}.$$

Formula 5(5)

The standard deviation of the juvenile homicide arrest data is

$$s = \sqrt{234.25} = 15.31.$$

The square root transformation restores the original units; we are back to *arrests* now and have solved the problem of *squared arrests*. Note that the standard deviation can never be negative because it is based on the variance, which is always a positive number.

Standard deviation: Computed as the square root of the variance, a measure of dispersion that is the mean of the deviation scores.

Substantively interpreted, the standard deviation is the mean of the deviation scores. In other words, it is the mean distance between the individual raw scores and the distribution mean. It indicates the general spread of the data by conveying information as to whether the raw values cluster close to the mean (thereby producing a relatively small standard deviation) or are more dispersed (producing a relatively large standard deviation). The standard deviation is generally presented in conjunction with the mean in the description of a continuous variable. We can say, then, that in 2009, Northeastern states had a mean of 12.00 juvenile arrests for homicides, with a standard deviation of 15.31 arrests. Figure 5.3 on page 86 displays the relationship between the mean and standard deviation pictorially for populations and samples.

RESEARCH EXAMPLE 5.1

Means and Standard Deviations in Criminal Justice/Criminology Research

a. Does the South have a culture of honor that increases gun violence?

Scholars have frequently noted that the South leads the nation in rates of violence and that gun violence is particularly prevalent in this region. This has led to the formation and proposal of multiple theories attempting to explain Southern states' disproportionate

(Continued)

(Continued)

involvement in gun violence. One of these theories is the "culture of honor" thesis that predicts that white male Southerners are more likely than their counterparts in other regions of the country to react violently when they feel that they have been disrespected or that they, their family, or their property have been threatened. Copes, Kovandzic, Miller, and Williamson (2009) tested this theory using data from a large, nationally representative survey of adults' reported gun ownership and defensive gun use (the use of a gun to ward off a perceived attacker). A primary independent variable was whether respondents themselves currently lived in the South. The main dependent variable was the number of times respondents had used a firearm (either fired or merely brandished) to defend themselves or their property against a perceived human threat within the past 5 years. The researchers reported the following descriptive statistics:

Variable	Mean	Standard Deviation	Percentage
DV: Number of defensive gun uses in the past five years	1.04	.19	
Percent of sample currently living in the South			34.0
Percentage of white state population born in the South	70.88	12.85	
White homicide rates in cities of respondents' residence	5.60	3.45	
Respondents' perceptions of crime in their neighborhoods (5–point scale)	2.50	1.12	

Source: Adapted From Table 1 in Copes, Kovandzic, Miller, and Williamson (2009).

The authors ran a statistical analysis to determine if living in the South or in a state where the majority of the population was born in the South was related to defensive gun use. They found that it was not: Neither currently living in the South nor living in a state populated primarily with Southerners increased the likelihood that respondents had used a gun defensively in the past five years. These findings refuted the Southern culture of honor thesis by suggesting that Southern white males are no more likely than white males in other areas of the country to resort to firearms to defend themselves.

b. Why does punishment often increase—rather than reduce—criminal offending?

Deterrence is perhaps the most pervasive and ingrained punishment philosophy in the United States and, indeed, in much of the world. It is commonly assumed that punishing someone for a criminal transgression will lessen the likelihood of that person reoffending in the future. Several studies have noted, however, that offending actually *increases* after someone has been punished. Offenders who have been caught, moreover, tend to believe that their likelihood of being caught again are very low because they think that is improbable that the same event would happen to them twice. This flawed probabilistic reasoning is called the "gambler's bias."

Pogarsky and Piquero (2003) attempted to learn more about the way that the gambler's bias may distort people's perceptions of the likelihood of getting caught for criminal wrongdoing. They distributed a survey with a drunk driving scenario to a sample of university students and asked the respondents about two dependent variables: (a) on a scale of 0 – 100, how likely they would be to drive under the influence in this hypothetical situation; and (b) on a scale of 0 – 100, how likely it is that they would be caught by police if they did drive while intoxicated. The researchers also gathered data on several independent variables, such as respondents' criminal histories (which were used to create a risk index scale), levels of impulsivity in decision making, and ability to correctly identify the probability of a flipped coin landing on tails after having landed heads side up four times in a row. The coin-flip question tapped into respondents' ability to use probabilistic reasoning correctly; those who said that the coin is more likely to land tails up were coded as engaging in the type of logical fallacy embodied by the gambler's bias. The researchers obtained the following means and standard deviations.

Variable	Mean	Standard Deviation
DV: Offending likelihood	38.28	36.14
DV: Perceived certainty of punishment	28.48	36.13
Number times pulled over while driving drunk	.24	.63
Number of days in past month consumed 3+ drinks	4.69	5.00
Number of times shoplifted	3.69	11.74
Number of times vandalized property	3.78	4.46
Number of 10 closest friends who have driven drunk at least once	4.50	3.34
Total risk index score	4.84	2.67

Source: Adapted From Appendix in Pogarsky and Piquero (2003).

(Continued)

(Continued)

Pogarsky and Piquero divided the sample into those at high risk of offending and those at low risk, and then analyzed the relationship between the gambler's fallacy and perceived certainty of punishment within each group. They found that the high-risk respondents' perceptions of certainty were not affected by the gambler's bias; even though 26 percent of people in this group did engage in flawed assessments of probabilities, this fallacious reasoning did not impact respondents' perceptions of the certainty of punishment.

Among low-risk respondents, however, there was a tendency for those who engaged in flawed probabilistic reasoning to believe that they stood a very low chance of detection relative to those low-risk respondents who accurately assessed the probability of a coin flip landing tails-side-up. The researchers concluded that people who are at high risk of offending may not even stop to ponder their likelihood of being caught and will proceed with a criminal act when they feel so inclined. Those at low risk, on the other hand, attempt to utilize probabilities to predict their chances of being apprehended and punished. The gambler's fallacy, therefore, may operate only among relatively naïve offenders who attempt to use probabilistic reasoning when making a decision about whether or not to commit a criminal offense.

A fundamental characteristic of any normal distribution, such as those displayed in Figure 5.4, is that approximately two thirds of the scores in the distribution lie within one standard deviation below and one standard deviation above the mean. This distance between one standard deviation below and one standard deviation above the mean constitutes the "normal" or "typical" range in a distribution. In a distribution of height, for instance, approximately two thirds of the sample would cluster fairly closely to the mean, while the remaining one third of the sample would be atypically tall or unusually short. These extreme values would indeed occur, but with relatively low frequency. People, objects, and places tend to cluster around their group means, with extreme values being infrequent and improbable. Remember this point; we will come back to it in later chapters.

Figure 5.3   Means and Standard Deviations for Normally Distributed Continuous Variables

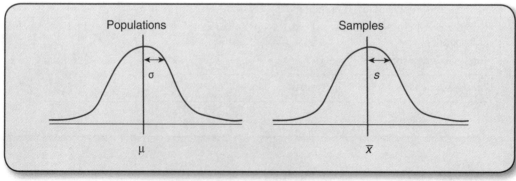

Figure 5.4 In a Normal Distribution, Approximately Two Thirds of the Scores Lie Within One Standard Deviation of the Mean

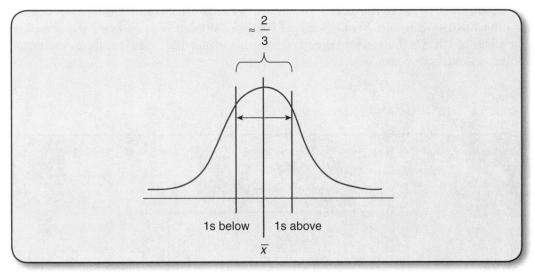

▣ SPSS

SPSS offers ranges, variances, and standard deviations. It will not provide you with variation ratios, but those are easy to calculate by hand. The process for obtaining measures of dispersion in SPSS is very similar to that for measures of central tendency. The juvenile homicide data set from Table 5.4 will be used to illustrate SPSS. First, use the *Analyze → Descriptive Statistics → Frequencies* sequence to produce the main dialog box. Then click the *Statistics* button in the upper right to produce the box displayed in Figure 5.5.

Figure 5.5 Using SPSS to Obtain Measures of Dispersion

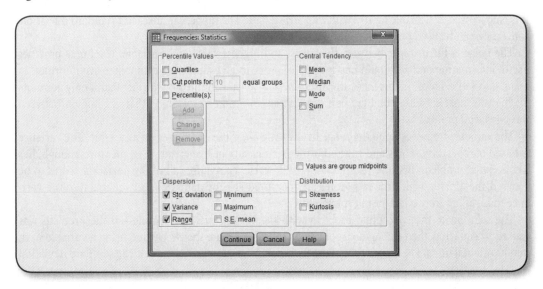

The range, variance, and standard deviation options are located in the lower left corner; click to insert a checkmark next to each, then click *Continue* and *OK*. The output in Figure 5.6 will be produced.

You can see in Figure 5.6 that all of the numbers generated by SPSS match those that we obtained by hand. This data set is called *Juvenile Arrests for Homicide, Northeast States 2009.sav* and is located on the companion website (**http://www.sagepub.com/gau**). Open this data file and run these measures of dispersion to replicate Figure 5.6.

Figure 5.6 SPSS Output for Measures of Dispersion

Statistics

Number of juvenile arrests for homicide in 2009

N	Valid	9
	Missing	0
Std. Deviation		15.30523
Variance		234.250
Range		36.00

CHAPTER SUMMARY

This chapter introduced four measures of dispersion: the variation ratio, range, variance, and standard deviation. The *VR* tells you the proportion of cases not located in the modal category. The *VR* can be computed on data of any level of measurement; however, it is the only measure of dispersion covered in this chapter that is available for use with categorical data. The range, variance, and standard deviation, conversely, can be used only with continuous variables.

The range is the distance between the lowest and highest value on a variable. The range provides useful information and so should be reported in order to give audiences comprehensive information about a variable; however, this measure's usefulness is severely limited by the fact that it only accounts for the two most extreme numbers in a distribution. It ignores everything that is going on between those two end points.

The variance improves upon the range by utilizing all of the raw scores on a variable. The variance is based on deviation scores and is an informative measure of dispersion. This measure, though, has a conceptual problem: Because computation of the variance requires all of the deviation scores to be squared, the units in which the raw scores are measured also, by definition, get squared too. The variance thus suffers from a lack of interpretability.

The solution to the conceptual problem with the variance is forthcoming—if the problem was caused by squaring, then the square root can be taken to negate the effects and bring a variable back into its original units. The square root of the variance is the standard deviation. The standard deviation is the mean of the deviation scores. Raw scores have a mean, and so do deviation scores. The former is simply called the mean, and the latter is the standard deviation.

The mean and standard deviation together are a staple set of descriptive statistics for continuous variables. The standard deviation will be key to many of the concepts that we will be discussing in the next few chapters. Approximately two thirds of the cases in any normal distribution are located between one standard deviation below and one standard deviation above the mean. These scores are within the normal or typical range; scores that are more than one standard deviation below or above the mean are relatively uncommon.

CHAPTER 5 REVIEW PROBLEMS

1. Compute the variation ratio for the following variable.

   Sex of Respondents in the National Crime Victimization Survey, 2004

Sex	f
Male	6,944
Female	8,179
	$N = 15,123$

2. Compute the variation ratio for the following variable.

   Annual Household Income of Respondents in the Police-Public Contact Survey, 2005

Income	f
Less than $20,000	30,791
$20,000 – $49,000	19,559
$50,000 or more	29,887
	$N = 80,237$

3. Compute the variation ratio for the following variable.

   Conviction Status of Jail Inmates in the Census of Jail Inmates, 2006

Status	f
Convicted	440,873
Not convicted, awaiting trial	270,712
	$N = 760,795$

4. Compute the variation ratio for the following variable.

Age of Stopped Drivers in the Police–Public Contact Survey, 2005

Age	f
16 – 19	5,618
20 – 29	12,513
30 – 39	14,336
40 – 49	16,452
50 – 59	13,643
60+	17,675
	$N = 80,237$

5. Compute the variation ratio for the following variable.

Race of Stopped Drivers in the Police–Public Contact Survey, 2005

Race	f
White	56,198
Black	8,677
Hispanic	10,540
Other	4,822
	$N = 80,237$

6. The table below contains UCR data showing the percentage of murders in each region of the country that were committed with firearms in 2009. (Note that this is a category combining all gun types.) Calculate the range.

Percent of Homicides Committed With Firearms in 2009, by Region

Region	Percent
Northeast	64.80
Midwest	69.40
South	68.20
West	64.60
$N = 4$	

7. Using the data in Question 6, calculate the mean.

8. Using the data in Question 6, calculate the variance.

9. Using the data in Question 6, calculate the standard deviation.

10. The table on the next page shows the proportion of murders in 2009 that were committed with knives and other cutting instruments. Calculate the range.

Proportion of Homicides Committed With Knives and Other
Cutting Instruments in a Random Sample of States, 2009

State	Proportion
Illinois	.08
Vermont	.57
Montana	.14
Iowa	.24
South Dakota	.45
Connecticut	.16
Mississippi	.15
$N = 7$	

11. Using the data in Question 10, calculate the mean.

12. Using the data in Question 10, calculate the variance.

13. Using the data in Question 10, calculate the standard deviation.

14. The following table contains the number of murders in Southern states that were committed with handguns in 2009. (No data were reported for Florida.) Calculate the range.

Number of Murders Committed With Handguns in Southern States, 2009 (UCR)

State	Number
Alabama	196
Arkansas	54
Delaware	20
Georgia	323
Kentucky	90
Louisiana	330
Maryland	297
Mississippi	83
North Carolina	243
Oklahoma	104
South Carolina	115
Tennessee	200
Texas	661
Virginia	108
West Virginia	20

15. Using the data in Question 14, calculate the mean.

16. Using the data in Question 14, calculate the variance.

17. Using the data in Question 14, calculate the standard deviation.

18. Explain the conceptual problem with the variance that is the reason why the standard deviation is generally used instead.

19. Explain the concept behind the standard deviation; that is, what does the standard deviation represent substantively?

20. Approximately ___ of any normally distributed set of data points lies between one standard deviation above and one standard deviation below the mean.

21. There is an SPSS file called *Florida LEMAS Data.sav* located on the accompanying (**http://www.sagepub.com/gau**). This file contains data from the Law Enforcement Management and Administrative Statistics survey, and the included agencies are the sample of municipal police departments in Florida that participated in the survey. One of the variables is called *female* and is the proportion of each department's total number of sworn full-time officers that is female. Using the *Analyze* → *Descriptive Statistics* → *Frequencies* menu, compute measures of central tendency *and* measures of variability for the variable *female*. As extra practice, try this out with the other variables in the data set.

KEY TERMS

Dispersion	Bounding rule	Range
Variation ratio	Rule of the complement	Standard deviation

GLOSSARY OF SYMBOLS INTRODUCED IN THIS CHAPTER

VR	Variation ratio
R	Range
s^2	Variance
s	Standard deviation

REFERENCES

Copes, H., Kovandzic, T. V., Miller, J. M., & Williamson, L. (2009). The lost cause? Examining the Southern culture of honor through defensive gun use. *Crime & Delinquency*. Prepublished August 12, 2009. DOI: 10.1177/0011128709343145

Pogarsky, G., & Piquero, A. R. (2003). Can punishment encourage offending? Investigating the "resetting" effect. *Journal of Research in Crime and Delinquency, 40*(1), 95–12.

PART II

PROBABILITY AND DISTRIBUTIONS

Part I of this book introdu[...] e mathematical and conceptual underp[...] nd other statistics that describe va[...] searchers want to do more than merely des[...] ze relationships between two or more vari[...] riptive statistic in the sample data. There is[...] which it was drawn—for reasons that wil[...] nnot simply be generalized to populations.[...] sample of people, for instance, and find a [...] mean is also 2.19. You may have drawn a s[...] tion.

Inferential statistics (a.k.a. hyp[...] earch-ers to bridge the gap between a des[...] e overarch-ing population. The purpose of infere[...] stic to be used in a manner such that the researcher can d[...] pulation. This procedure is grounded in **probability theory**. Probab[...] oundation for statistical tests and is therefore the subject of Part II of this bo[...] Part I as having established the founda-tional mathematical and formulaic concept[...] y for inferential tests and of Part II as laying out the theory behind the strategic use of those d[...]scriptive statistics.

Inferential Statistics: The field of statistics in which a descriptive statistic derived from a sample is employed probabilistically to make a generalization or inference about the population from which the sample was drawn.

Probability Theory: Logical premises that form a set of predictions about the likelihood of certain events or the empirical results that one would expect to see in an infinite set of trials.

So far, you've survived 100% of your worst days - you're doing great ♥

Part II is heavily grounded in theory. You will not see SPSS sections or much use of research examples. This is because probability is implicit—rather than explicit—in criminal justice/criminology research; probability is the proverbial "person behind the curtain" who is pulling the levers and making the machine run but who usually remains hidden from view. While there will be formulas and calculations, your primary task in Part II is to form conceptual understandings of the topics presented. When you understand the logic behind inferential statistics, you will be ready for Part III.

Chapter 6

Probability

A **probability** is the likelihood that a particular event will occur. You use probabilistic reasoning on a daily basis. Your use of probabilistic reasoning is evident in your musings over whether or not to buy a lottery ticket, the chances that you will get a raise or promotion if you perform a work task exceptionally well, and the likelihood that you will get pulled over if you drive five miles per hour over the speed limit. These mental exercises all involve predictions about the likelihood that a certain event will (or will not) occur.

Probability: The likelihood that a certain event will occur.

Probabilities are linked intricately with proportions; in fact, the probability formula is basically a spinoff of Formula 3(1). Flip back to this formula right now for a refresher. Let us call a particular event of interest A. Events are phenomena of interest that are being studied. The probability that event A will occur can be symbolized as $p(A)$ (pronounced p of A) and written formulaically as

$$p(A) = \frac{\text{The number of times event } A \text{ can occur}}{\text{The number of outcomes possible}}.$$

 Formula 6(1)

Coin flips are the classic example of probabilities. Any two-sided, fair coin may land on either heads or tails when flipped, so the denominator in the probability formula is 2. There is only one tail side on a coin, so the numerator is 1. Probabilities, like proportions, are always expressed as decimals, so the fraction must be divided out. The probability of the coin landing tails side up, then, is

$$p(\textit{tails}) = \frac{1}{2} = .50.$$

Any time you flip a fair coin, there is a .50 probability that the coin will land on tails. Of course, the probability that it will land heads side up is also .50. Note that the two probabilities together sum to 1.00; that is,

$$p(tails) + p(heads) = .50 + .50 = 1.00.$$

The probabilities sum to 1.00 because heads and tails are the only two possible results and thus constitute an exhaustive list of outcomes. The coin, moreover, must land (it will not simply hover in midair above your head when you toss it upward), so the probability sums to 1.00 because the landing of the coin is inevitable.

Think about rolling one fair die. A die has six sides, so any given side has probability $\frac{1}{6} = .17$ of being the side that lands face up. If you were asked, "What is the probability of obtaining a 3 on a single roll of a fair die?" your answer would be ".17." There are 52 cards in a standard deck and only one of each number and suit, so if a card is randomly selected from that deck, each card has a $\frac{1}{52} = .02$ probability of being the card that is selected. If someone asked you, "What is the probability of selecting the two of hearts from a full deck?" you would respond, ".02."

Table 6.1   Gender of PPCS Respondents

Sex	f	p
Male	38,078	.47
Female	42,159	.53
	N = 80,237	Σ = 1.00

The major difference between proportions and probabilities is that proportions are purely descriptive, while probabilities represent predictions. Consider Table 6.1, which is adapted from Table 3.2 in Chapter 3. The table shows the proportion of the PPCS sample that is male and the proportion that is female. If you threw all 80,237 PPCS respondents into a gigantic hat and randomly drew one, what is the probability that the person you selected would be female? .53! The probability that the person would be male? .47. This is the relationship between proportions and probabilities.

THE PROBABILITY THAT CRIMINAL OFFENSES WILL RESULT IN THE ARREST OF A SUSPECT

Consider this rather morbid question: If you committed a criminal offense, how likely is it that you would be arrested for it? Write down your expectations about this probability. Feel free to break it down by offense type. Then take a look at the following bar graph showing the proportion of offenses cleared by arrest or exceptional means in 2009, as reported by the Uniform Crime Reports. These proportions indicate that the probability of clearance is actually quite low, especially for property crimes—the likelihood that a burglary would be cleared in 2009 is .13! How do these probabilities compare to the one(s) you wrote down? Are you surprised?

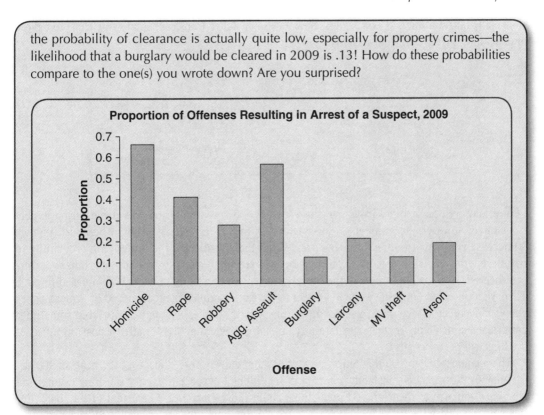

Probabilities are based on an infinite number of trials; in other words, probabilities usually concern what is expected over the long run. Think back to the coin flip example. Since *p*(*tails*) on any given flip is .50, then over the course of a long-term flipping session, exactly half of the flips should yield tails.

Theoretical Prediction: A prediction, grounded in logic, about whether or not a certain event will occur.

Empirical Outcome: A numerical result from a sample, such as a mean or frequency. Also called observed outcomes.

There is a distinct difference, though, between **theoretical predictions** and **empirical outcomes** (also called *observed outcomes*). Quite simply, what you expect is not always what you get. You can see this in practice. First, imagine that you flipped a coin six times. How many of those flips would you expect to land tails side up? Three, of course, because every flip has *p*(*tails*) = .50 and so you would expect half of all flips to result in tails. Now, grab a real coin, flip it six times, and record each outcome.

How many tails did you get? Try 20 flips. Now how many tails? Transfer the following chart to a piece of paper and fill it in with your numbers.

6 flips	proportion tails $= \dfrac{?}{6} =$
20 flips	proportion tails $= \dfrac{?}{20} =$

You may well have found in your 6-flip exercise that the number of tails departed noticeably from three; however, the number of tails yielded in the 20-flip exercise should be approximately 10 because you increased the number of trials and thereby allowed the underlying probability to manifest in the empirical outcomes. The knowledge that half of coin flips will produce tails over time is a theoretical prediction, whereas the flip experiment you just conducted is an empirical finding. Theory guides us in outlining what we expect to see (for example, if you are asked how many times you would expect to see tails in a series of six flips, your best guess would be three), but sometimes empirical results do not match expectations (for example, you may see no tails, or possibly all six of the flips will produce tails).

The relationship between theoretical predictions and empirical findings is at the heart of statistical analysis. Researchers constantly compare observations to expectations to determine if empirical outcomes conform to theory-based predictions about those outcomes. The extent of the match (or mismatch) between theory and reality leads to conclusions about the accuracy of those theories.

In every type of research study and every sample that is drawn, there are various arrays of probabilities. If, for instance, in studying sentencing you took a sample of four people who had been convicted of felonies and counted the ones who received prison sentences, the possible empirical outcomes that you might see are that zero, one, two, three, or all four received a prison sentence. Each of these potential outcomes (0, 1, 2, 3, and 4) has a particular probability of being *the* one that actually takes place. Some outcomes have large probabilities (meaning they are likely to occur) and some have small probabilities (meaning they are unlikely to happen).

STUDY TIP: THOUGHT EXERCISES

If you are thus far finding the material to be about as clear as mud, then you are experiencing the perfectly normal learning process for this topic. These ideas take time to sink in. A highly recommended study technique is to spend time simply thinking about these concepts. Work them over in your mind. It may be advisable to take fairly frequent breaks as you read so that you can digest what you have read so far before continuing on.

A **probability distribution** is a table or graph showing the full array of theoretical probabilities for any given variable. These probabilities represent not what we *actually* see but, rather, the gamut of potential empirical outcomes and each outcome's probability of being the one that actually happens. Probability distributions are theoretical, not empirical. A probability distribution is constructed on the basis of an underlying parameter or statistic (such as a proportion or a mean) and represents the probability associated with each possible outcome. Two types of probability distributions are discussed in this chapter: binomial and continuous.

Probability Distribution: A table or graph showing the entire set of probabilities associated with every possible empirical outcome.

▣ DISCRETE PROBABILITY: THE BINOMIAL PROBABILITY DISTRIBUTION

A **trial** is a particular act with multiple different possible outcomes (for example, rolling a die, where the die will land on any one of its six sides). **Binomials** are trials that have exactly two possible outcomes (this type of trial is also frequently called *dichotomous* or *binary*). Coin flips are binomials because coins have two sides. Research Example 6 describes two types of binomials that criminal justice/criminology researchers have examined.

Trial: An act that has several different possible outcomes.

Binomial: A trial with exactly two possible outcomes. Also called a dichotomous or binary variable.

RESEARCH EXAMPLE 6.1

Dichotomous Outcomes in Criminal Justice/Criminology Research

Are police officers less likely to arrest an assault suspect when the suspect and the alleged victim are intimate partners? Critics of the police response to intimate partner violence have accused police of being "soft" on offenders who abuse intimates. Klinger (1995) used a variable measuring whether or not police made an arrest when responding to an assault of any type. The variable was coded as *arrest/no arrest*. He then examined whether

(Continued)

(Continued)

the probability of arrest was lower when the perpetrator and victim were intimates as compared to assaults between strangers or nonintimate acquaintances. The results indicated that police were unlikely to make arrests in *all* types of assault, regardless of the victim-offender relationship, and that they were not less likely to arrest offenders who victimized intimate partners relative to those who victimized strangers or acquaintances.

Are minority drivers more likely than white drivers to be stopped by police? Among those motorists stopped, are persons of color more likely than whites to be asked for consent to search their vehicles? The racial profiling debate is not just about whether police single out minority drivers for disproportionately intense traffic enforcement—it extends to the question of pretext stops. Pretext stops occur when police pull a vehicle over for a minor traffic violation with the express intent of having face-to-face contact with the driver and asking for consent to search the vehicle. These types of stops, though extremely controversial, are legal pursuant to the U.S. Supreme Court's decision in *Whren v. United States* (1996). So, do police single out minority drivers for stops and searches? No single study can answer this question definitively, but Smith and Petrocelli (2001) sought to contribute useful information to the debate. They analyzed the race of stopped drivers and, moreover, they examined a subsample of drivers who had been searched in the course of the traffic stop to determine if driver race predicted whether a search was a consent search or some other type of search (for example, incident to arrest). The dependent variable, then, was search type and was measured as consent search or other search. Contrary to what would be expected from a racial profiling perspective, white drivers' searches were significantly more likely than minority drivers' to be conducted pursuant to consent. The researchers offered several potential explanations for this finding. First, police may actually be more likely to seek consent of white drivers. A second possible reason is that white drivers may be more likely than drivers of color to give consent when it is requested. Finally, the majority of drivers arrested during the course of traffic stops are nonwhite, which may have confounded the analysis by reducing the number of minority drivers available for consent requests.

The binomial probability distribution is a list of expected probabilities; it contains all possible results over a set of trials and lists the probability of each result. Continuing the imprisonment study example from above, you would think of each person's sentencing hearing as a trial with two possible outcomes: incarceration or nonincarceration. As described above, there are five possible outcomes (ranging from none incarcerated to all four incarcerated). Each of these possible outcomes has a particular probability of being the one that occurs in a particular sample of offenders, and the binomial probability distribution tells you what those probabilities are so that you can predict what you would expect to see in any given sample of offenders.

So, how do we go about building the binomial probability distribution? The distribution is constructed using the **binomial coefficient**. The formula for this coefficient is a bit intimidating, but each component of the coefficient will be discussed individually in the following pages so that when you are done reading, you will understand the coefficient and how to use it. The binomial coefficient is given by the formula:

$$p(r) = \binom{N}{r} p^r q^{n-r}, \text{ where}$$

Formula 6(2)

$p(r)$ = the probability of r,

r = the number of successes,

N = the number of trials/sample size,

p = the probability that a given event will occur,

q = the probability that a given event will not occur.

The ultimate goal of binomials is to find the value of $p(r)$ for every possible value of r. The resulting list of $p(r)$ values is the binomial probability distribution.

Let us break the formula down into its constituent parts using an example. Suppose you are studying pretrial release. You gather a random sample of five criminal defendants awaiting disposition on felony charges. Trials here are the pretrial release decisions. They have two possible outcomes (*released, not released*) and are thus binomials. Suppose you find that three of the five were released. You want to draw a conclusion about this result; specifically, you want to know whether this finding is typical or, conversely, whether it is unusual and contrary to theoretical expectations. You can find this out by constructing a binomial probability distribution to determine the probability associated with each possible outcome. You can then compare the probabilities to figure out whether it is likely or unlikely that three out of five defendants would be released.

Successes and Sample Size: *N* and *r*

The binomial coefficient formula contains the variables N and r. N represents sample size or the total number of trials. In the present example, five defendants were randomly selected, so $N = 5$. The symbol r represents the number of **successes** or, in other words, the number of times the outcome of interest happens over N trials. In the current example, *success* is defined as a defendant obtaining pretrial release. If we are interested in the probability that three defendants will have been released, then $r = 3$.

Success: The outcome of interest in a trial.

The Number of Ways r Can Occur, Given N: The Combination

If a release is symbolized r, then for the purpose of an illustration, let us call a detention d. Consider the following sets of possible arrangements of outcomes:

$$\{r,d,d,d,d\} \{d,r,d,d,d\} \{d,d,r,d,d\} \{d,d,d,r,d\} \{d,d,d,d,r\}$$

What these sets tell you is that there are five different ways for $r = 1$ (that is, one success) to occur over $N = 5$ trials. Defendant 1 might be the success and Defendants 2, 3, 4, and 5 the failures, or it is possible that Defendant 5 is the success and Defendants 1, 2, 3, and 4 are the failures, and so on. The same holds true for any number of successes—there are many different ways that $r = 2$, $r = 3$, and so on all the way up to $r = 5$ can occur in terms of which defendants are the successes and which are the failures.

The total number of ways that r can occur in a sample of size N is called a **combination** and is calculated as

$$\frac{N!}{r!(N-r)!}, \text{ where}$$

N = the total number of trials or total sample size,

r = the number of successes,

$!$ = factorial.

The factorial symbol $!$ tells you to multiply together the series of descending whole numbers starting with the number to the left of the symbol all the way down to 1.00. If $r = 3$, then $r! = 3 \cdot 2 \cdot 1 = 6$. If $N = 5$, then $N! = 5 \cdot 4 \cdot 3 \cdot 2 \cdot 1 = 120$.

There is also shorthand notation for the combination formula that saves us from having to write the whole thing out. The shorthand notation is $\binom{N}{r}$, which is pronounced "N choose r." Be very careful! This is not a fraction. Do not mistake it for "N divided by r."

STUDY TIP: USING A CALCULATOR TO COMPUTE FACTORIALS AND COMBINATIONS

Most calculators meant for use in math classes (which excludes the calculator that came free with that new wallet you bought) will compute factorials and combinations for you. The factorial function is labeled $!$ The combination formula is usually represented by nCr or just C. Depending on the type of calculator you have, these functions are probably either located on the face of the calculator and accessed using a *2nd* or *alpha* key, or can be found in a menu accessed using the *math, stat,* or *prob* buttons, or a certain combination of these functions. Take a few minutes right now to find these functions on your calculator.

$$1! = 1$$

$$0! = 1$$

The combination formula can be used to replicate the above longhand demonstration wherein we were interested in the number of ways that one release and four detentions can occur. Plugging the numbers into the combination formula yields

$$\binom{5}{1} = \frac{5!}{1!(5-1)!} = \frac{120}{1(4!)} = \frac{120}{1(24)} = 5$$

It is thus confirmed that there are five different ways for one person to be released and the remaining four to be detained.

In our hypothetical study sample of five defendants, $r = 3$. How many combinations of three are there in a sample of five? Plug the numbers into the combination formula to find out:

$$\binom{5}{3} = \frac{5!}{3!(5-3)!} = \frac{120}{3!(2!)} = \frac{120}{6(2)} = \frac{120}{12} = 10$$

There are 10 combinations of three in a sample of five. In the context of the present example, there are 10 different ways for three defendants to be released and two to be detained. This has to be accounted for in the computation of the probability of each possible result, which is why you see the combination formula included in the binomial coefficient.

The Probability of Success and the Probability of Failure: *p* and *q*

The probability of success (symbolized p) is at the heart of the binomial probability distribution. This number is obtained on the basis of prior knowledge or theory. In the present example pertaining to pre-trial release, we can use information from the Bureau of Justice Statistics to find a value of p. According to the *Felony Defendants in Large Urban Counties, 2006* report, 58% of felony defendants in the study year obtained some form of pretrial release. Let us say, then, that each defendant's probability of being released pending trial is $p = .58$.

We know that release is not the only possible outcome, though—defendants can be detained, too. Because we are dealing with events that have two potential outcomes, we need to know the probability of **failure** in addition to that of success. The probability of failure is represented by the letter q; to compute the value of q, the rule of the complement must be invoked. In Chapter 5's discussion of the variation ratio, you learned that the probability of the event of interest *not* occurring is equal to 1.00 minus the probability that it *will* occur; that is:

$$p(Not\,A) = 1.00 - p(A).$$

What we are doing now is exactly the same thing, with the small change that $p(Not\ A)$ will now be represented by the letter q. So,

$$q = 1.00 - p.$$

Formula 6(4)

The probability that a felony defendant will be detained prior to trial (that is, will *not* obtain pretrial release) is

$$q = 1.00 - .58 = .42.$$

Putting IT ALL Together: Using the Binomial Coefficient to Construct the Binomial Probability Distribution

Using p and q, the probability of various combinations of successes and failures can be computed. When there are r successes over N trials, then there are $N - r$ failures over that same set of trials. There is a formula called the **restricted multiplication rule for independent events** that guarantees that the probability of r successes and $N - r$ failures is the product of p and q. In the present example, there are three successes (each with probability $p = .58$) and two failures (each with probability $q = .42$). You should also recall from prior math classes that exponents can be uses as shorthand for multiplication when a particular number is multiplied by itself many times. Finally, we also have to account for the combination of N and r. The probability of three successes is thus:

$$p(3) = \binom{5}{3} \cdot p(success) \cdot p(success) \cdot p(success) \cdot p(failure) \cdot p(failure)$$

$$= \binom{5}{3} \cdot p \cdot p \cdot p \cdot q \cdot q = \binom{5}{3} p^3 q^2 = \binom{5}{3}(.58^3)(.42^2) = 10(.20)(.18) = .36.$$

Failure: Any outcome other than success or the event of interest.

Restricted Multiplication Rule for Independent Events: A rule of multiplication that allows the probability that two events will both occur to be calculated as the product of each event's probability of occurrence: that is, $p(A\ and\ B) = p(A) \cdot p(B)$.

Given a population probability of .58, there is a .36 probability that in a sample of five defendants, three would obtain pretrial release.

This procedure must be repeated for every possible value of r in order to create the binomial probability distribution. The most straightforward way to do this is to construct a table like Table 6.2. Every row in the table uses a different r value, while the values of N, p, and q are fixed. The rightmost column, $p(r)$, is obtained by multiplying the $\binom{N}{r}$, p^r, and q^{N-r} terms.

Table 6.2   The Binomial Probability Distribution for Pretrial Release Among Five Felony Defendants

r	$\binom{N}{r}$	p^r	q^{N-r}	$p(r)$
0	$\binom{5}{0} = 1$	$.58^0 = 1.00$	$.42^{5-0=5} = .01$	.01
1	$\binom{5}{1} = 5$	$.58^1 = .58$	$.42^{5-1=4} = .03$	.09
2	$\binom{5}{2} = 10$	$.58^2 = .34$	$.42^{5-2=3} = .07$	.24
3	$\binom{5}{3} = 10$	$.58^3 = .20$	$.42^{5-3=2} = .18$	.36
4	$\binom{5}{4} = 5$	$.58^4 = .11$	$.42^{5-4=1} = .42$	.23
5	$\binom{5}{5} = 1$	$.58^5 = .07$	$.42^{5-5=0} = 1.00$	.07

The values in the $p(r)$ column tell you the probability of each potential outcome r being the one that occurs in any given random sample of five defendants. The probability that two of the five defendants will be released pending trial is .24, while the probability that all five will be released is .07. We can use this column to answer the original question regarding the probability that in a random sample of five defendants, three would have been released while awaiting trial. You can see from the table that the probability of $r = 3$ is $p(3) = .36$. This, as it happens, is the highest probability in the table, meaning that it is exactly the outcome that we would most expect to see in any given random sample. Our results are quite typical! Another way to think of it is that given the underlying population probably of .58, the results we found here are not unusual in any way but are, rather, right in line with what we would expect to see.

Contrast this to the conclusion we would draw if none of the five defendants in the sample had been released. This is an extremely improbable event with only $p(0) = .01$ likelihood of occurring. It would be rather surprising to find this empirical result, and it might lead us to wonder if there was something unusual about our sample. We might investigate the possibility that the county we drew the sample from had particularly strict policies regulating pretrial release, or that we happened to draw a sample of defendants charged with especially serious crimes.

There are two neat and important things about the binomial probability distribution. The first is that it can be graphed using a bar chart (Figure 6.1) so as to form a visual display of the numbers in Table 6.2. The horizontal axis contains the *r* values and the bar height is determined by $p(r)$. The bar chart makes it easy to determine at a glance which outcomes are most and least likely to occur.

Figure 6.1   The Binominal Probability Distribution for Pretrial Release in a Sample of Five Defendants

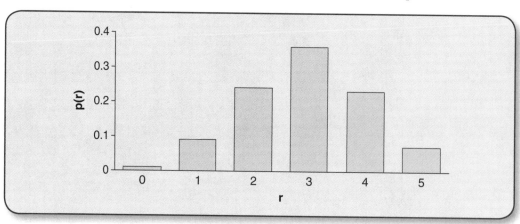

The second important thing about the binomial probability distribution is that the $p(r)$ column sums to 1.00. This is because the binomial distribution is exhaustive; that is, all possible values of *r* are included in it. Any time an exhaustive list of probabilities is summed, the result will be 1.00. Memorize this point! It is applicable in the context of continuous probability distributions, too, and will be revisited shortly.

▣ CONTINUOUS PROBABILITY: THE STANDARD NORMAL CURVE

The binomial probability distribution, as you know by now, is applicable for trials with two potential outcomes; however, criminal justice and criminology researchers often use continuous variables. The binomial probability distribution has no applicability in this context. Continuous variables are represented by a distribution called the **normal curve**. Figure 6.2 shows this curve's familiar bell shape.

It is a *unimodal, symmetric* curve with an area of 1.00. It is unimodal because it peaks once and only once. It is symmetric because the two halves (split by the mean) are identical to one another. It has an area of 1.00 because it encompasses all possible values of the variable in question. Just as the sum of the binomial probability distribution's $p(r)$ column always sums to 1.00 because all values that *r* could possibly take on are contained within the table, so the normal curve's tails stretch out to negative and positive infinity. This may sound impossible, but remember that this is a *theoretical* distribution. This curve is built on probabilities, not actual data.

Normal curve: A distribution of raw scores from a sample or population that is symmetric, unimodal, and has an area of 1.00. Normal curves differ from one another in metrics, means, and standard deviations.

Figure 6.2   The Normal Curve for Continuous Variables

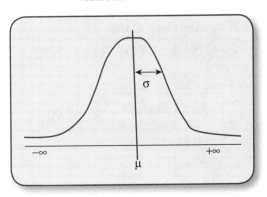

The characteristics that determine a normal curve's location on the number line and its shape are its mean and standard deviation, respectively. When expressed in raw units, normal curves are scattered about the number line and take on a variety of shapes. This is a product of variation in metrics, means, and standard deviations. Figure 6.3 depicts this concept.

Figure 6.3   Variation in Normal Curves

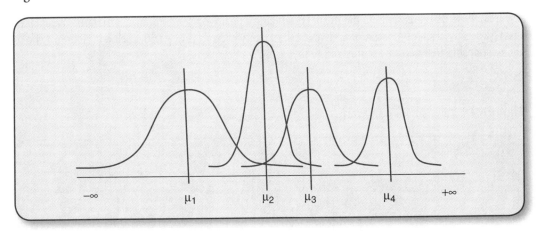

This inconsistency in locations and dimensions of normal curves can pose a problem in statistics. It is impossible to determine the probability of a certain empirical result when every curve differs from the rest. What is needed is a way to standardize the normal curve so that all variables can be represented by a single curve. This widely applicable single curve is constructed by converting all of a distribution's raw scores to *z* **scores.** The *z* score transformation is straightforward:

$$z_x = \frac{x - \bar{x}}{s}, \text{ where}$$

Formula 6(5)

z_x = the *z* score for a given raw score *x*,

x = a given raw score,

$\bar{x}$ = the distribution mean,

s = the distribution standard deviation.

---

z score: A standardized version of a raw score that offers two pieces of information about the raw score: (1) how close it is to the distribution mean; and (2) whether it is greater than or less than the mean.

A *z* score conveys two pieces of information about the raw score upon which the *z* score is based. First, the absolute value of the *z* score reveals the location of the raw score in relation to the distribution mean. *Z* scores are expressed in standard deviation units. A *z* score of .25, for example, tells you that the underlying raw score is exactly one fourth of one standard deviation away from the mean. A *z* score of −1.50, likewise, signifies a raw score that is one and one half standard deviations away from the mean.

The second piece of information is given by the sign of the *z* score. Although standard deviations are always positive, *z* scores can be negative. A *z* score's sign indicates whether the raw score that that *z* score represents is greater than the mean (producing a positive *z* score) or is less than the mean (producing a negative *z* score). A *z* score of .25 is above the mean, while a score of −1.50 is below it. Figure 6.4 shows the relationship between raw scores and their *z* score counterparts.

Figure 6.4 Raw Scores and *z* Scores

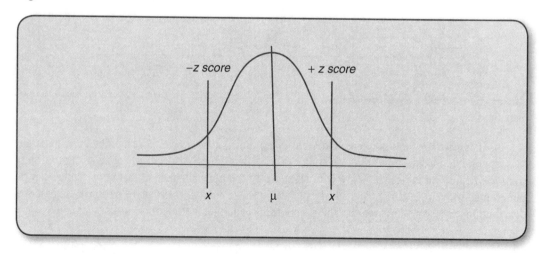

When all of the raw scores in a distribution have been transformed to *z* scores and plotted, the result is the **standard normal curve**. The *z* score transformation dispenses with the original, raw values of a variable and replaces them with numbers representing their position *relative to* the distribution mean. A normal curve, then, is a curve constructed of raw scores, while the standard normal curve is composed of *z* scores.

Standard Normal Curve: A distribution of *z* scores. The curve is symmetric, unimodal, has a mean of zero, a standard deviation of 1.00, and an area of 1.00.

Like ordinary normal curves, the standard normal curve is symmetric, unimodal, and has an area of 1.00. Unlike regular normal curves, though, the standard normal curve's mean and standard deviation are fixed at 0 and 1, respectively. They remain fixed irrespective of the units in which the variable is measured or the original distribution's mean and standard deviation. This allows probabilities to be computed.

To understand the process of using the standard normal curve to find probabilities, it is necessary to comprehend that in this curve, area is the same as proportion and probability. A given area of the curve (such as the area between two raw scores) also represents the proportion of scores that are between those two raw values. This is, in turn, also equal to the probability that any randomly selected raw score will be located between these two scores. Figure 6.5 displays the relationship between *z* scores and areas.

In Chapter 5, you learned that approximately two thirds of the scores in any normal distribution lie between one standard deviation below and one standard deviation above the mean for that set of scores. (Refer back to Figure 5.4.) In Figure 6.5, you can see that .3413 (or 34.13%) of the scores are located between the mean and a one standard deviation distance from the mean. If you add the proportion of cases—that is, area of the curve—that is one standard deviation below the mean to that proportion or area that is one standard deviation above, you get .3413 + .3413 = .6826, or 68.26%. This is just over two thirds! The bulk of scores in a normal distribution cluster fairly closely to the center and those scores that are within one standard deviation of the mean (i.e., *z* scores that have absolute values of 1.00 or less) are considered the typical or *normal* scores.

Z scores that are greater than 1.00 or less than −1.00 are relatively rare, and they get increasingly rare as you trace the number line away from zero in either direction. These very large *z* scores do happen, but they are improbable, and some of them are incredibly unlikely.

This is all getting very abstract, so let us get an example going. As the technology for DNA testing becomes more available and as the number of postconviction exonerations based on DNA testing climbs disturbingly, police, prosecutors, and defense attorneys are seeking pretrial DNA testing to identify perpetrators faster and to bolster confidence in the accuracy of convictions. According to the 2005 Census of Publicly Funded Forensic Crime Laboratories (CPFFCL; see Data Sources 6.1), state crime labs received a mean of 610 requests for forensic DNA analyses during this year. The standard deviation was 581.

Figure 6.5   Area Under the Standard Normal Curve

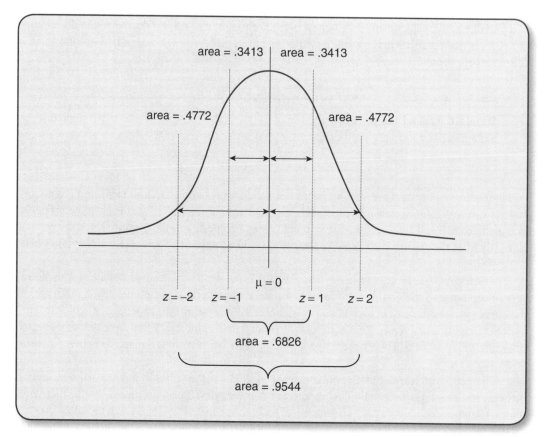

DATA SOURCES 6.1

THE CENSUS OF PUBLICLY FUNDED
FORENSIC CRIME LABORATORIES

The Census of Publicly Funded Forensic Crime Laboratories is conducted periodically by the Bureau of Justice Statistics. All forensic crime labs operated at the state, federal, local, or other level of the public sector are contacted and asked to complete a survey about their operations (CPFFCL Codebook, 2008). Questionnaire items pertain to issues such as budgets, staffing, funding sources, salaries, total requests received, and total requests completed.

The Illinois State Police crime lab received 475 DNA requests in 2005. To find this lab's z score, plug the numbers into Formula 6(5):

$$z_{475} = \frac{475 - 610}{581} = \frac{-135}{581} = -.23.$$

We can tell from the Illinois lab's z score that the number of DNA analysis requests this lab received was .23 (or just over one fifth) of one standard deviation below the nationwide mean, thus putting the Illinois crime lab squarely within the normal zone on the curve. Note that we would not have to know either the mean or Illinois's raw score to be able to interpret this z score and get a sense as to how this lab compares to the others in the nation; the z score alone gives us this information.

Now for another one. The New Jersey State Police DNA lab received 1,853 requests, making its z score

$$z_{1853} = \frac{1853 - 610}{581} = \frac{1243}{581} = 2.14.$$

The New Jersey lab's score is just over two standard deviations above the mean. Is this score inside or outside of the normal or typical range on the curve? Outside! This score well surpasses a score of 1.00, making it an atypically high score.

z scores can be used to find areas, proportions, or probabilities; remember that all three of these are the same (on the standard normal curve) under the curve. To do this, the **z table** is used. The z table is a chart containing the area of the curve that is between the mean and a given z score. Appendix B contains the z table. The area associated with a particular z score is found by decomposing the score into an *x.x* and *.0x* format such that the first half of the decomposed score contains the digit and the number in the 10ths position, and the second half contains a zero in the 10ths place and the number that is in the 100ths position. For Illinois, the z score would be broken down as $-.23 = -0.2 + (-.03)$. Note that it does not matter that the z score is negative because the standard normal curve is symmetric and, therefore, the z table is used the same way irrespective of an individual score's sign. Go to the z table and locate the .2 row and .03 column, then trace them to their intersection. The area is .0910. Figure 6.4 depicts this concept. Thinking in terms of proportions, if the entire curve's area is 1.00, then .0910 or 9.10% of that curve is between the mean and a z-score of −.23. This is not very much! Since this score is less than one quarter of one standard deviation away from the mean, there is very little area in this space.

---

z table: A table containing a list of z scores and the area of the curve that is between the distribution mean and each individual z score.

Finding the area between the mean and a z score is informative, but what is generally of interest to criminology/criminal justice researchers is the area *beyond* that z score; that is, researchers usually want to know how big or small a z score would have to be in order to fall into the very tip of either the positive or negative tail of the distribution. These scores have a very low probability of occurring, so when they do happen, it is a pretty big deal.

Let us find the area beyond z for the New Jersey lab. That lab's raw score was 1,853 and its z score was 2.14. The fact that it is positive means that this score is on the right-hand side of the standard normal curve. For purposes of the z table, $2.14 = 2.1 + .04$. The area between the mean and that z score is .4838 (see Figure 6.6).

This, though, is not the area we are looking for—we want to know the area in the *remainder* of the right-hand side of the curve. This number can be derived using subtraction. You already learned that the total area under the curve is 1.00, and you know that the curve is symmetric as split in half by the mean. The area in each half of the distribution, then, is .50. In the present example, the area between the mean and z is known (.4838), as is the total area on that side of the curve (.50), so the final unknown piece—the area beyond z—can be found by subtraction, as such:

$$.50 - .4838 = .0162.$$

Thus, .0162 (or approximately .02, if we round to two decimal places) of the scores in the standard normal distribution for DNA requests lies beyond the raw value of 1,853. This is a very small proportion! Scores of 1,853 and greater are extremely rare and unlikely events.

Figure 6.6   The Area in the Tail Beyond a Z Score of 2.14

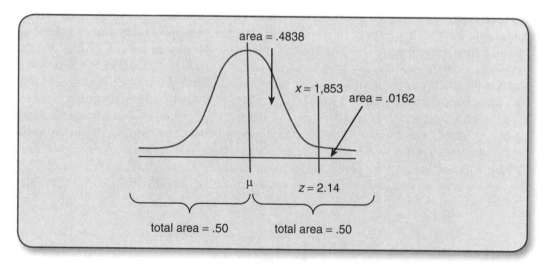

Another useful feature of the standard normal curve is that just as z scores can be used to find areas, areas can likewise be used to find z scores. This requires using the z table backward. We might, for instance, want to know the z scores that lie in the upper 5% (.05) or 1% (.01) of the distribution. These are very unlikely values and are of interest because of their relative rarity. How large would a z score have to be such that only .05 or .01 of scores is above it?

The process of finding these z scores employs similar logic about the area of each side of the curve. First, we know that the z table provides the area between z and the mean; it does not tell you about the

Figure 6.7 Using Areas to Find *Z* Scores

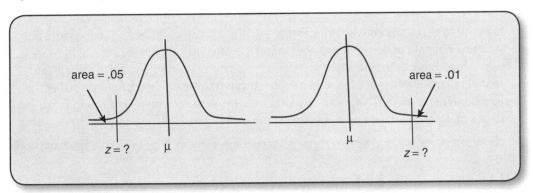

area beyond *z*. This problem is surmounted using subtraction. Let us start with the area of .01. The area between the mean and *z* = .50 − .01 = .49 (Figure 6.7).

The second step is to scan the areas listed in the body of the *z* table to find the one that is *closest to .49*. The closest area is .4901. Third, find the *z* score associated with the identified area. The area of .4901 is in the *2.3* row and *.03* column, so the *z* score is 2.3 + .03 = 2.33. And that is the answer! We now know that .01, or 1%, of scores in the standard normal distribution area have *z* scores greater than 2.33.

Now let us find the *z* score associated with an area of .05, but with a twist: We will place this area in the lower (left-hand) tail of the distribution. Recall that we were working in the upper (right-hand) tail when we did the above example using an area of .01. The first and second steps of the process are the same no matter which side of the distribution is being analyzed. Subtraction shows that .50 − .05 = .45 of the scores is between the mean and *z*. Going to the table, you can see that there are actually two areas that are closest to .45: .4495 and .4505. The *z* score for the former is 1.64 and that for the latter is 1.65; *however*, since we are on the left (or negative) side of the curve, these *z* scores are actually −1.64 and −1.65. You must be aware of the sign of your *z* score! *Z* scores on the left side of the standard normal curve are negative.

Finally, since there are two *z* scores in this instance, they must be averaged:

$$z = \frac{-1.64 + (-1.65)}{2} = \frac{-3.29}{2} = -1.645 \approx -1.65.$$

That is the answer! *Z* scores less than −1.65 have a .05 or less probability of occurring. In other words, these extremely small scores happen only 5% of the time or less. That is unlikely indeed.

STUDY TIP: NO NEGATIVE AREAS

It is imperative to understand that although *z* scores can be negative, areas are always positive. When you are working on the left side of the standard normal *z* curve, scores will be negative but areas will not be.

STUDY TIP: *Z* SCORES AND PROBABILITIES

It is easy to get confused when discussing z scores and probabilities or areas under the standard normal curve, but remember that these are two ways of expressing the exact same idea. Large z scores are associated with small probabilities, and small z scores represent large probabilities. The larger the absolute value of a z score, the smaller the likelihood of observing that score. Z scores near zero are not unlikely or unusual, while scores that are very far away from zero are relatively rare.

CHAPTER SUMMARY

Probability is the basis of inferential statistics. This chapter introduced two of the major probability distributions: binomial and standard normal. The binomial distribution is for categorical variables that have two potential outcomes (dichotomous or binary variables), and the standard normal curve applies to continuous variables. The binomial probability distribution is constructed on the basis of some underlying probability that is derived from research or theory. The standard normal curve consists of z scores, which are scores associated with known areas or probabilities. Raw scores can be transformed to z scores using a simple conversion formula. Because the area under the standard normal curve is a constant 1.00, areas can be added and subtracted, thus allowing probabilities to be determined on the basis of z scores and vice versa.

Both distributions are theoretical, which means that they are constructed upon the basis of logic and mathematical theory. They can be contrasted to empirical distributions, which are distributions made from actual, observed raw scores in a sample or population. Empirical distributions are tangible; they can be manipulated and analyzed. Theoretical distributions exist only in the abstract.

Theoretical distributions give you information about the probabilities of certain empirical phenomena. When a researcher is looking at an empirical result, what that researcher wants to know is how typical or atypical this empirical finding is. Is it a routine, typical finding, or is it an unusual and highly improbable result? These questions are the building blocks of inferential statistics.

CHAPTER 6 REVIEW PROBLEMS

1. Eight police officers are being randomly assigned to two-person teams.
   a. Identify the value of *N*.
   b. Identify the value of *r*.
   c. How many combinations of *r* are possible in this scenario? Do the combination by hand first, and then check your answer using the combination function on your calculator.

2. Nine jail inmates are being randomly assigned to three-person cells.
   a. Identify the value of *N*.
   b. Identify the value of *r*.
   c. How many combinations of *r* are possible in this scenario? Do the combination by hand first, and then check your answer using the combination function on your calculator.

3. In a sample of seven parolees, three are rearrested within two years of release.

    a. Identify the value of N.
    b. Identify the value of r.
    c. How many combinations of r are possible in this scenario? Do the combination by hand first, and then check your answer using the combination function on your calculator.

4. Out of six persons recently convicted of felonies, five are sentenced to prison.

    a. Identify the value of N.
    b. Identify the value of r.
    c. How many combinations of r are possible in this scenario? Do the combination by hand first, and then check your answer using the combination function on your calculator.

5. Four judges in a sample of eight do not believe that the law provides them with sufficient sanction options when sentencing persons convicted of crimes.

    a. Identify the value of N.
    b. Identify the value of r.
    c. How many combinations of r are possible in this scenario? Do the combination by hand first, and then check your answer using the combination function on your calculator.

    For each of the following variables, identify the distribution—either binomial or standard normal—that would be the appropriate theoretical probability distribution to represent that variable. Remember that this is based on the variable's level of measurement.

6. Defendants' completion of a drug court program, measured as *success* or *failure*.

7. The total lifetime number of times someone has been arrested.

8. Crime victims' reporting of their victimization to police, measured as *reported* or *did not report*.

9. The number of months of probation received by juveniles adjudicated guilty on delinquency charges.

10. City crime rates.

11. Prosecutorial charging decisions, measured as *filed charges* or *did not file charges*.

12. According to the Uniform Crime Reports (UCR), 64% (or .64) of aggravated assaults reported to police in 2009 were cleared by arrest. Take this as your value of p and do the following:

    a. Compute the binomial probability distribution for a random sample of five aggravated assaults, with r being defined as the number of assaults that are cleared.
    b. Based on the distribution, what is the outcome you would most expect to see?
    c. Based on the distribution, what is the outcome you would least expect to see?
    d. What is the probability that three or fewer aggravated assaults would be cleared?
    e. What is the probability that four or more would be cleared?

13. According to the Uniform Crime Reports (UCR), 70% or .70 of persons convicted of felonies were sentenced to a term of incarceration (either prison or jail). Take this as your value of p and do the following:

    a. Compute the binomial probability distribution for a random sample of six aggravated assaults, with r being defined as the number of assaults that are cleared.
    b. Based on the distribution, what is the outcome you would most expect to see?

   c. Based on the distribution, what is the outcome you would least expect to see?

   d. What is the probability that two or fewer of the convicted felons would be incarcerated?

   e. What is the probability that five or more of them would be incarcerated?

14. According to the UCR, 49% or .49 of the hate crimes that were reported to police in 2009 were racially motivated. Take this as your value of p and do the following:

   a. Compute the binomial probability distribution for a random sample of six hate crimes, with r being defined as the number that are racially motivated.

   b. Based on the distribution, what is the outcome you would most expect to see?

   c. Based on the distribution, what is the outcome you would least expect to see?

   d. What is the probability that two or fewer of the hate crimes were motivated by race?

   e. What is the probability that three or more were racially motivated?

15. According to the UCR, 59% (.59) of murders in 2009 were committed with firearms. Take this as your value of p and do the following:

   a. Compute the binomial probability distribution for a random sample of five murders, with r being defined as the number that are committed with firearms.

   b. Based on the distribution, what is the outcome you would most expect to see?

   c. Based on the distribution, what is the outcome you would least expect to see?

   d. What is the probability that one or fewer murders were committed with firearms?

   e. What is the probability that four or more murders were committed with firearms?

The 2003 Law Enforcement and Management Statistics (LEMAS) survey reported that the mean number of municipal police per 1,000 citizens in U.S. cities with populations of 100,000 or more was 1.98 (s = .85). Use this information to answer the following questions.

16. Los Angeles, California, had 1.16 city police per 1,000 citizens.

   a. Convert this raw score to a z score.

   b. Find the area between the mean and z.

   c. Find the area in the tail of the distribution beyond z.

17. New York City, New York, had 4.60 police per 1,000 residents.

   a. Convert this raw score to a z score.

   b. Find the area between the mean and z.

   c. Find the area in the tail of the distribution beyond z.

18. Birmingham, Alabama, had 3.41 police per 1,000 residents.

   a. Convert this raw score to a z score.

   b. Find the area between the mean and z

   c. Find the area in the tail of the distribution beyond z.

19. Seattle, Washington, had 2.18 police per 1,000 residents.

   a. Convert this raw score to a z score.

   b. Find the area between the mean and z.

   c. Find the area in the tail of the distribution beyond z.

Find the z scores associated with the areas given below. Remember to add a negative sign to the score when necessary.

20. What *z* scores fall into the upper .15 of the distribution?

21. What *z* scores fall into the upper .03 of the distribution?

22. What *z* scores fall into the lower .02 of the distribution?

23. What *z* scores fall into the upper .10 of the distribution?

24. What *z* scores fall into the lower .005 of the distribution?

25. What *z* scores fall into the lower .10 of the distribution?

KEY TERMS

Inferential statistics

Probability theory

Probability

Theoretical prediction

Empirical outcome

Probability distribution

Binomial

Normal curve

Standard normal curve

Z table

GLOSSARY OF SYMBOLS INTRODUCED IN THIS CHAPTER

p	The probability of success
N	The total number of trials; the sample size
r	The number of successes over a set of N trials
$\binom{N}{r}$	The combination of "N choose r"; the number of ways for r successes to happen over N trials
q	The probability of failure
z	A score expressed in standard deviation units

REFERENCES

Klinger, D. A. (1995). Policing spousal assault. *Journal of Research in Crime and Delinquency, 32*(2), 308–324.

Smith, M. R., & Petrocelli, M. (2001). Racial profiling? A multivariate analysis of police traffic stop data. *Police Quarterly, 4*(1), 4–27.

Whren v. United States, 517 U.S. 806 (1996).

Chapter 7

Population, Sample, and Sampling Distributions

A population is the entire universe of objects, people, places, or other units of analysis that a researcher wishes to study. Criminal justice/criminology researchers use all manner of populations. Bouffard (2010), for instance, examined the relationship between men's military service during the Vietnam era and their criminal offending later in life. Kane and Cronin (2010) attempted to determine whether there was a relationship between order maintenance policing and violent crime in communities. Morris and Worrall (2010) investigated whether prison architectural design influenced inmate misconduct. These are three examples of populations—male Vietnam veterans, communities, prison inmates—that can form the basis for study.

The problem is that populations are usually far too large for researchers to examine directly. There are millions of men in the United States, thousands of communities nationwide, and hundreds of thousands of inmates who engage in misconduct. Nobody can possibly study any of these populations in its entirety. Samples are thus drawn from populations of interest. Samples are subsets of populations. Morris and Worrall (2010), for example, drew a random sample of 2,500 inmates. This sample, unlike its overarching population, was of manageable size and could be analyzed directly.

Populations and samples give rise to three types of distributions: population, sample, and sampling. A **population distribution** contains all values in the entire population, while a **sample distribution** shows the shape and form of the values in a sample pulled from a population. Population and sample distributions are both empirical. They are made of raw scores derived from actual people or objects. **Sampling distributions**, by contrast, are theoretical arrays of sample statistics. Each of these is discussed in turn below.

Population distribution: An empirical distribution made of raw scores from a population.

Sample distribution: An empirical distribution made of raw scores from a sample.

Sampling distribution: A theoretical distribution made out of an infinite number of sample statistics.

▣ EMPIRICAL DISTRIBUTIONS: POPULATION AND SAMPLE DISTRIBUTIONS

Population and sample distributions are both empirical because they exist in reality; every person or object in the population or sample has a value on a given variable that can be measured and plotted on a graph. To illustrate these two types of distributions, we can use the BJS 2005 Census of State and Federal Adult Correctional Facilities (see Data Sources 7.1). This data set contains information on all adult correctional facilities that house state and federal prisoners. No sampling was done; every adult correctional facility in the U.S. was asked to provide information. This makes the CSFACF a population data set.

DATA SOURCES 7.1

THE CENSUS OF STATE AND FEDERAL ADULT CORRECTIONAL FACILITIES (CSFACF)

The CSFACF is a BJS-sponsored effort to gather and maintain data on adult correctional facilities owned or operated by state and federal governments. All U.S. federal and state departments of corrections are asked to participate, and most comply; Illinois was the only jurisdiction that refused to provide correctional data for the 2005 survey (Stephan, 2005). The data set contains information such as facility security level; total prisoner population and population broken down by age, gender, and race; staffing levels; whether or not the facility was under court order during the year that the survey was conducted; and the types of vocational and educational programming offered to inmates. Many of the prisons in the sample are privately owned and operated, which allows for comparisons between privately and publicly operated correctional facilities.

Figure 7.1 displays the population distribution for the variable *total number of inmates*, which is a measure of the number of prisoners housed in each facility at year-end 2005. Every facility's inmate count was plotted to form this histogram. You should be able to recognize immediately that this distribution has an extreme positive skew.

What might the distribution look like for a sample pulled from this population? The SPSS program can be commanded to select a random sample of cases from a data set, so this function will be used to simulate sampling. A random sample of 300 facilities was pulled from the CSFACF data file and the variable *total number of inmates* plotted to form Figure 7.2. The sample distribution looks similar to the population distribution, which makes sense because this is a random sample of fairly large size ($N = 300$). The sample is thus a miniature version of the population.

Figure 7.1   Population Distribution for Total Inmates per Correctional Facility ($N = 1,817$)

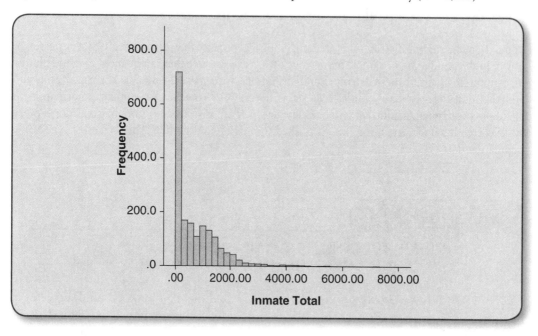

Figure 7.2   Sample Distribution ($N = 300$)

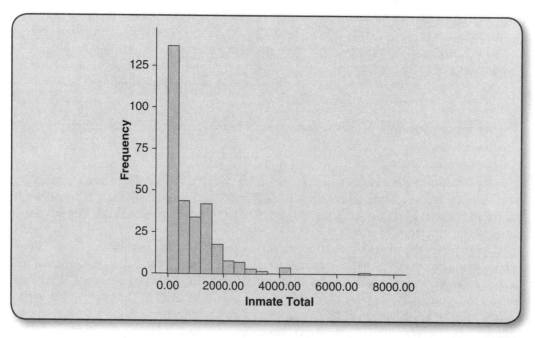

STUDY TIP: STATISTICS AND PARAMETERS

Remember the difference between statistics and parameters. Statistics are numbers derived from samples, while parameters are numbers based on populations. Statistics are estimates of parameters. We will be using the follow two sets of statistics and parameters very frequently throughout this and later chapters.

	Sample Statistic	*Population Parameter*
Mean	$\bar{x}$	μ
Standard deviation	s	σ
Proportion	$\hat{p}$	P

So, what criminal justice/criminology researchers usually *want* is to make a statement about a population, but what they actually *have* in front of them to work with is a sample. This creates a conundrum because sample statistics cannot simply be generalized to population parameters. Statistics are estimates of population parameters and, moreover, they are estimates that contain error. Population parameters are fixed. This means that they have only one mean and standard deviation. Sample statistics, by contrast, vary from sample to sample because of **sampling error**. Sampling error arises from the fact that infinite random samples can be drawn from any population. Any given sample that a researcher *actually* draws is only one of a multitude that he or she *could* have drawn. Figure 7.3 depicts this. Every potential sample has its own distribution and set of descriptive statistics. In any given sample, these statistics might be exactly equal to, roughly equal to, or completely different from their corresponding parameters.

Sampling error: The uncertainty introduced into a sample statistic by the fact that any given sample is only one of an infinite number of samples that could have been drawn from that population.

The CSFACF can be used to illustrate the effects of sampling error because, as described above, this data set is a population rather than a sample. We will continue using the variable *total number of inmates housed*. The population mean is $\mu = 785.53$, and the standard deviation is $\sigma = 935.20$. Drawing five random samples of 300 facilities each and computing the mean and standard deviation of each sample produces the data in Table 7.1.

Look how the means and standard deviations vary—this is sampling error! Here, we have the benefit of knowing the true population mean and standard deviation, but that is usually not the case

Figure 7.3   An Infinite Number of Random Samples Can Be Drawn From Any Population

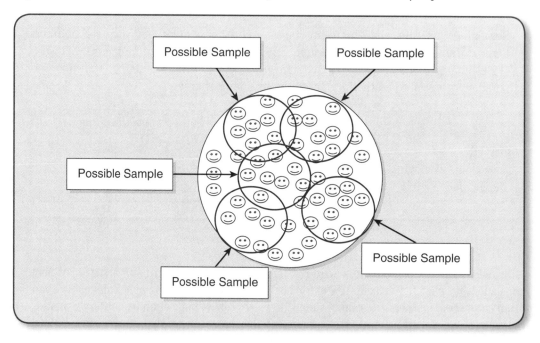

Table 7.1   Means and Standard Deviations for the Total Inmates in the Population ($N = 1,821$) and Five Random Samples of ($N = 300$ in Each)

	Population	Sample 1	Sample 2	Sample 3	Sample 4	Sample 5
Mean	$\mu = 785.53$	806.66	888.66	772.98	751.16	739.04
Std. Dev.	$\sigma = 935.20$	919.68	1008.97	919.77	903.45	840.20

in criminal justice/criminology research. What researchers generally have is one sample and no information about the population as a whole. Imagine, for instance, that you drew Sample 3 in Table 7. This sample's mean and standard deviation are reasonable approximations of—though clearly not equivalent to—their corresponding population values, but you would not know that. Now picture Sample 2 being the sample you pulled for a particular study. This mean and standard deviation are markedly discrepant from the population parameters, but, again, you would be unaware of that.

There is, thus, a chasm between samples and populations that is created by sampling error and prevents inferences from being made directly. It would be a mistake to draw a sample, compute its mean, and automatically assume that the population mean must be equal or close to the sample mean. What is needed is a bridge between samples and populations so that inferences can be reliably drawn. This bridge is the sampling distribution.

▣ THEORETICAL DISTRIBUTIONS: SAMPLING DISTRIBUTIONS

Sampling distributions, unlike population and sample distributions, are theoretical; that is, they do not exist as empirical realities. We have already worked with theoretical distributions in the form of the binomial and standard normal distributions. Sampling distributions are theoretical because they are based on the notion of infinite samples being drawn from a single population. What sets sampling distributions apart from empirical distributions is that sampling distributions are created not from raw scores but, rather, from sample statistics. These statistics can be means, proportions, or any other statistic that can be calculated on a sample. Imagine plotting the means in Table 7 to form a histogram, like Figure 7.4.

Figure 7.4 Histogram of the Sample Means in Table 7.1

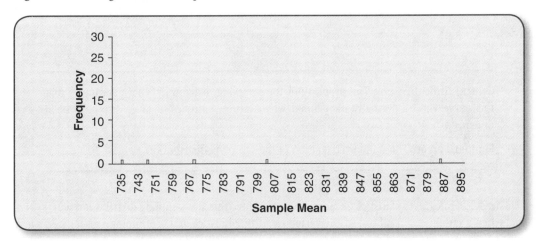

Not terribly impressive, is it? Definitely leaves something to be desired. That is because there are only five samples. Sampling distributions start to take shape only when many samples have been drawn. If we continued the iterative process of drawing a sample, computing the mean, plotting that mean, throwing the sample back, and pulling a new sample, the distribution Figure 7.4 would gradually start looking something like the curve in Figure 7.5.

Now the distribution has some shape! It looks much better. It is, moreover, not just any old shape—it is a normal curve. What you have just seen is the **central limit theorem** (CLT) in action. The CLT states that any time descriptive statistics are computed from an infinite number of large samples, the resulting sampling distribution will be normally distributed. The sampling distribution clusters around the true population mean (here, 785.53) and if you were to compute the mean of the sampling distribution (i.e., the mean of means; symbolized $\mu_{\bar{x}}$), the answer you obtained would match the true population mean. Its standard deviation (called the **standard error**, symbolized $\sigma_{\bar{x}}$) is smaller than the population standard deviation because there is less dispersion, or variability, in means than in raw scores. This produces a narrower distribution, particularly as the size of the samples increases.

Figure 7.5 Sampling Distribution of Means for Total Inmate Count

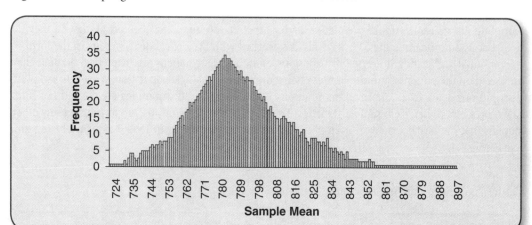

Central limit theorem: The property of the sampling distribution that guarantees that this curve will be normally distributed when infinite samples of large size have been drawn.

Standard error: The standard deviation of the sampling distribution.

The CLT is integral to statistics because of its guarantee that the sampling distribution will be normal when a sample is large. Most of the variables criminal justice and criminology researchers work with are skewed. The central limit theorem saves the day by ensuring that even skewed variables will produce normal sampling distributions. The inmate count variable demonstrates this. Compare Figure 7.5 to Figure 7.2. Figure 7.2 is highly skewed, yet the sampling distribution in Figure 7.5 is normal. That is because even when raw values produce severe skew, sample statistics will still hover around the population parameter and fall symmetrically on each side of it. Some statistics will be greater than the parameter and some will be smaller, but the majority will be approximately or exactly equal to it.

The key benefit of the sampling distribution being normally distributed is that the standard normal curve can be used. Everything we did in Chapter 6 with respect to using raw scores to find z scores, z scores to find areas, and areas to find z scores can be done with the sampling distribution. The applicability of z, though, is contingent upon N being large.

All descriptive statistics have sampling distributions to which the CLT applies. In Chapter 6, you learned that nationwide, .58 (or 58%) of felony defendants are granted pretrial release. Figure 7.6 sketches what the sampling distribution of proportions for the pretrial release variable might look like.

Sample standard deviations also have sampling distributions. Interestingly enough, there are even sampling distributions for differences between means and proportions. If you drew infinite random samples from not one but *two* populations, computed a mean or proportion each time, and

Figure 7.6   Sampling Distribution of Proportions

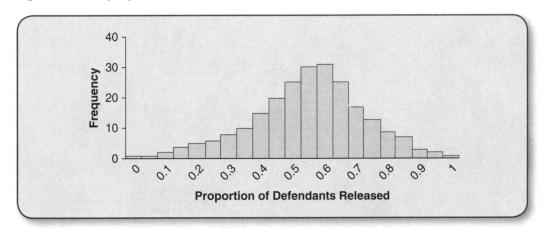

then subtracted one mean or proportion from the other and plotted that difference score, a normal curve would emerge. The sampling distribution of differences between means and proportions will be revisited in Chapter 11.

🔲 WHEN SAMPLE SIZE IS SMALL: THE *t* DISTRIBUTION

The central limit theorem allows us to assume that the sampling distribution is normal when sample size *N* is large. When this criterion is met, the *z* distribution can be used. But what is a "large" sample? And what happens when a sample is not large enough to permit use of the *z* distribution?

Large is a vague adjective in statistics because there is no formal rule specifying the dividing line between small and large samples. Generally speaking, large samples are those containing at least 100 cases. When $N \geq 100$, the sampling distribution can be assumed to be normally distributed.

It is generally not advisable to work with samples where $N \leq 99$, but there are times when it is unavoidable. In these situations, the *z* distribution cannot be employed and a different distribution is needed. The distribution researchers turn to is the ***t* distribution**. The *t* distribution—like the *z* curve—is symmetrical, unimodal, and has a constant area of 1.00. The key difference between the two is that *t* is a family of several different curves rather than one fixed, single curve like *z* is. The *t* distribution changes shape depending on the size of the sample. When the sample is small, the curve is wide and flattish; as the sample size increases, the *t* curve becomes more and more normal until it looks identical to the *z* curve. See Figure 7.7.

***t* distribution:** A family of curves whose shapes are determined by the size of the sample. All *t* curves are unimodal, symmetrical, and have an area of 1.00.

Figure 7.7   The Family of *t* curves

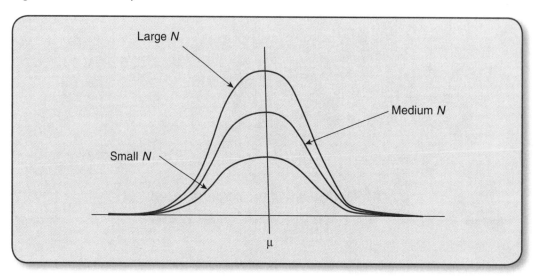

The *t* distribution's flexibility allows it to accommodate samples of various sizes. It is a very important theoretical probability distribution because it allows researchers to do much more than they would be able to if *z* were their only option. All else being equal, large samples are better than small ones, and it is always advisable to work with large samples when possible. When a small sample must be used, though, *t* is a trustworthy alternative.

CHAPTER SUMMARY

Criminal justice/criminology researchers seek information about populations. Populations, though, are usually too large to analyze directly. Samples are therefore pulled from them and statistical analyses are applied to these samples instead. Population and sample distributions are made of raw scores that have been plotted. These are empirical distributions.

Sampling error, though, introduces an element of uncertainty into sample statistics. Any given sample that is drawn is only one of a multitude of samples that could have been drawn. Because of sampling error, statistics are merely estimates of their corresponding population parameters and cannot be interpreted as matching them exactly. In this way, there is a gap between samples and populations.

The sampling distribution links samples to populations. Sampling distributions are theoretical curves made out of infinite sample statistics. All descriptive statistics have sampling distributions. The central limit theorem ensures that sampling distributions are normal when sample sizes are large (that is, $N \geq 100$). When this is the case, the *z* distribution can be used. When samples are small, though, the sampling distribution cannot be assumed to be normal. The *t* distribution solves this problem because *t* is a family of curves that change shape depending on sample size. The *t* distribution is more flexible than *z* and must be used any time $N \leq 99$.

CHAPTER 7 REVIEW PROBLEMS

1. Population distributions are

   a. empirical.
   b. theoretical.

2. Sample distributions are

   a. empirical.
   b. theoretical

3. Sampling distributions are

   a. empirical.
   b. theoretical.

4. The _____ distribution is made from the raw scores in a sample.

5. The _____ distribution is made from statistics calculated on infinite samples.

6. The _____ distribution is made from the raw scores in a population.

7. The central limit theorem guarantees that so long as certain conditions are met, a sampling distribution will be

   a. positively skewed.
   b. normally distributed.
   c. negatively skewed.

8. For the central limit theorem's promise of distribution shape to hold true, samples must be

   a. large.
   b. small.

9. When a sample contains 100 or more cases, the correct probability distribution to use is the _____ distribution.

10. When a sample contains fewer than 100 cases, the correct probability distribution to use is the _____ distribution.

11. A researcher gathers a sample of 200 people, asks each one how many times he or she has been arrested, and then plots each person's response. From the list below, select the type of distribution that this researcher has created.

    a. A sample distribution with a large N
    b. A population distribution with a small N
    c. A sampling distribution with a large N
    d. A sample distribution with a large N

12. A researcher gathers a sample of 49 police departments, finds out how many officers were fired for misconduct in each department over a two-year time span, and plots each department's score. From the list below, select the type of distribution that this researcher has created.

    a. A population distribution with a small N
    b. A population distribution with a large N

c. A sampling distribution with a small N

d. A sample distribution with a small N

13. A researcher gathers a sample of 20 cities, calculates each city's mean homicide rate, and plots that mean. Then the researcher puts that sample back into the population and draws a new sample of 20 cities and computes and plots the mean homicide rate. The researcher does this repeatedly. From the list below, select the type of distribution that this researcher has created.

    a. A sampling distribution with a large N
    b. A population distribution with a small N
    c. A sampling distribution with a small N
    d. A population distribution with a large N

14. A researcher has data on each of the nearly 2,000 adult correctional facilities in the United States and uses them to plot the number of inmate-on-inmate assaults that took place inside each prison in a one-year time span. From the list below, select the type of distribution that this researcher has created.

    a. A sample distribution with a large N
    b. A population distribution with a small N
    c. A sampling distribution with a large N
    d. A population distribution with a large N

15. A researcher gathers a sample of 132 people and computes the mean number of times the people in that sample have shoplifted. The researcher then puts this sample back into the population and draws a new sample of 132 people, for whom the researcher computes the mean number of times shoplifted. The researcher does this repeatedly. From the list below, select the type of distribution that this researcher has created.

    a. A sample distribution with a large N
    b. A sampling distribution with a large N
    c. A sample distribution with a small N
    d. A population distribution with a large N

KEY TERMS

Population distribution	Sampling error	t distribution
Sample distribution	Central limit theorem	
Sampling distribution	Standard error	

GLOSSARY OF SYMBOLS INTRODUCED IN THIS CHAPTER

$\mu_{\bar{x}}$	The mean of the sampling distribution
$\sigma_{\bar{x}}$	The standard deviation of the sampling distribution, called the standard error
t	A distribution that can accommodate sample sizes of fewer than 100 cases

REFERENCES

Bouffard, L. A. (2010). Period effects in the impact of Vietnam-era military service on crime over the life course. *Crime & Delinquency*. Prepublished September 8, 2010. DOI: 10.1177/0011128710372455

Kane, R. J., & Cronin, S. W. (2010). Associations between order maintenance policing and violent crime: Considering the mediating effects of residential context. *Crime & Delinquency*. Prepublished May 28, 2009. DOI: 10.1177/0011128709336940

Morris, R. G., & Worrall, J. L. (2010). Prison architecture and inmate misconduct: A multilevel assessment. *Crime & Delinquency*. Prepublished May 28, 2009. DOI: 10.1177/0011128709336940

Stephan, J. J. (2005). *Census of state and federal adult correctional facilities, 2005*. Washington, DC: U. S. Department of Justice, Bureau of Justice Statistics.

Chapter 8

Point Estimates and Confidence Intervals

A ny given sample that is drawn from a population is only one of a multitude of samples that *could have* been drawn. Every sample that is drawn (and every sample that could potentially be drawn) has its own descriptive statistics, such as a mean or proportion. This phenomenon is called **sampling error**. The variation in sample statistics such as means and proportions precludes direct inference from a sample to a population. It cannot be assumed that a mean or proportion in a sample is an exact match to the mean or proportion in the population because sometimes sample statistics are very similar to their corresponding population parameters, and sometimes they are quite dissimilar to these parameters. There is, then, always an element of uncertainty in a sample statistic, or what can also be called a **point estimate**.

Sampling error: The uncertainty introduced into a statistic by the fact that any sample that is drawn is one of infinite samples that could have been drawn, and that a sample statistic is therefore not necessarily equal to the population parameter.

Point estimate: A sample statistic, such as a mean or proportion.

Confidence interval: A range of values spanning a point estimate that is calculated so as to have a certain probability of containing the population parameter.

Fortunately, though, a procedure exists for calculating a range of values within which the parameter of interest may lie. This range stretches out on each side of the point estimate and is called a **confidence interval** (*CI*). The confidence interval acts as a sort of "bubble" that introduces flexibility into the estimate. It is much more likely that an estimate of the value of a population parameter is accurate when the estimate is a range of values rather than one single value.

Try thinking about it this way: Suppose I guessed that you are originally from Chicago. This is a very precise prediction! Of all the cities and towns in the world, I narrowed my guess down to a single area. Given its precision, though, this prediction is very likely to be wrong; there are 7 billion people in the world and only about 2.8 million of them live in Chicago. My "point estimate" (Chicago) is probably inaccurate.

But what if I instead guessed that you are from the state of Illinois? I have increased my chances of being correct because I have broadened the scope of my estimate. If I went up another step and predicted that you are from the Midwest—without specifying a city or state—I have further increased my probability of being right. It is still possible that I am wrong, of course, but I am far more likely to guess your place of origin correctly when I guess a large geographical area, such as a region, than when I guess a much smaller one, such as a city.

This is, conceptually, what a confidence interval does. It offers a "buffer zone" that allows for greater confidence in the accuracy of a prediction. It also allows us to determine the probability that the prediction we are making is accurate. Using distributions—specifically, the z and t probability curves—we can figure out how likely it is that our confidence interval truly does contain the true population parameter. The probability that the interval contains the parameter is called the **level of confidence**.

Level of confidence: The probability that a confidence interval contains the population parameter. Commonly set at 95% or 99%.

▣ THE LEVEL OF CONFIDENCE: THE PROBABILITY OF BEING CORRECT

In the construction of confidence intervals, you get to choose your level of confidence (that is, the probability that your confidence interval accurately estimates the population parameter). This may sound great at first blush—why not just choose 100% confidence and be done with it, right?—but confidence is actually the classic double-edged sword because there is a tradeoff between it and precision. Think back to the Chicago/Illinois/Midwest example. The Chicago guess has a very low probability of being correct (we could say that there is a low level of confidence in this prediction), but it has the benefit of being a very precise estimate because it is just one city. The Illinois guess carries an improvement in confidence because it is a bigger geographical territory; however, because it is bigger, it is also less precise. If I guess that you are from Illinois and I am right, I am still left with many unknown pieces of information about you. I would not know which part of the state you are from, whether you hail from a rural farming community or a large urban center, the socioeconomic characteristics of your place of origin, and so on.

The problem is further exacerbated if all I guess is that you are from the Midwest—now I would not even know which state you are from, much less which city! If I want to be 100% sure that I will guess your place of origin correctly, I have to put forth "planet Earth" as my prediction. That is a terrible estimate. If you want greater confidence in your estimate, then, you have to pay the price of reduced precision and, therefore, a diminished amount of useful information.

Confidence levels are expressed in percentages and can, strictly speaking, be any percentage between 0 and 99.$\bar{9}$. Although there is no "right" or "wrong" level of confidence in a technical sense, 95% and 99% have become conventional in criminal justice/criminology research. Because of the tradeoff between confidence and precision, a 99% CI has a greater chance than a 95% one of being correct, but the 99% one will be wider and less precise. A 95% CI will carry a slightly higher likelihood of error but will yield a more informative estimate. You should select your level of confidence by deciding whether it is more important that your estimate be correct or that it be precise.

Confidence levels are set a priori, which means that you must decide whether you are going to use 95% or 99% before you begin constructing the interval. The reason for this is that the level of confidence affects the calculation of the interval. You will see this when we get to the CI formula.

Since we are dealing with probabilities, we have to face the unpleasant reality that our prediction may be incorrect. The flipside of the probability of being right (that is, your confidence level) is the probability of being wrong. Consider:

$$100\% - 95\% = 5\%$$

$$100\% - 99\% = 1\%$$

Each of these traditional levels of confidence carries a corresponding probability that a confidence interval does *not* contain the true population parameter. If the 95% level is selected, then there is a 5% chance that the CI will not contain the parameter; a 99% level of confidence generates a 1% chance of an inaccurate CI.

You will, unfortunately, probably never know whether the sample you have in front of you is one of the 95% or 99% that is correct, or whether it is one of the 5% or 1% that is not. There is no way to tell; you just have to compute the confidence interval and hope for the best. This is an intractable problem in statistics and one that we all must accept.

Four types of confidence intervals will be discussed in this chapter: CIs for means with large samples ($N \geq 100$), for means with small samples ($N \leq 99$), for proportions or percentages with large samples, and for proportions or percentages with small samples. All four types of CIs are meant to improve the accuracy of point estimates by providing a range of values that most likely contains the true population parameter.

回 CONFIDENCE INTERVALS FOR MEANS WITH LARGE SAMPLES

When a sample is of large size ($N \geq 100$), the z distribution (the standard normal curve) can be used to construct a CI around a sample mean. Confidence intervals for means with large samples are computed as

$$\bar{x} \pm z_\alpha \left(\frac{s}{\sqrt{N-1}} \right), \text{ where}$$

Formula 8(1)

$\bar{x}$ = the sample mean,

z_α = the z score associated with a given alpha level,

α = the probability of being wrong,

s = the sample standard deviation,

N = the sample size.

The starting point, you can see, is the sample mean. A certain value will be added to and subtracted from the mean to form the interval. That value is the end result of the term on the right side of the $\pm$ operator. The z score's subscript α (this is the Greek letter *alpha*) represents the probability that the *CI* does not contain the true population parameter. Recall that every level of confidence carries a certain probability of inaccuracy; this probability of inaccuracy is α or, more formally, the **alpha level**. Alpha is computed as

$$\alpha = 1 - \text{confidence level.}$$

Alpha level: The opposite of the confidence level; that is, the probability that a confidence interval does not contain the true population parameter.

Critical value: The value of z or t associated with a given alpha level.

Two-tailed test: A statistical test in which alpha is split in half and placed into both tails of the z or t distribution.

Using Formula 8(2) for the conventional confidence levels that we are working with here,

$$\alpha_{95\%} = 1 - .95 = .05$$

$$\alpha_{99\%} = 1 - .99 = .01.$$

You will never enter α itself into the *CI* formula; rather, you use α to find the **critical value** of z and you enter that critical value into the formula (this is the z_α term). Since the z curve is a single, fixed distribution, the z score for an alpha of .05 will always be 1.96; likewise, the z score for an alpha of .01 will always be 2.58. Since we are only using 95% and 99% *CIs* in this chapter, these are the only two z scores you will be plugging into the formula for means with large samples.

There is another piece of information that you need to know about *CIs*. Confidence intervals are **two-tailed**. This is because the normal curve has two halves that are split by the mean, with the result being—to put a rather pessimistic spin on it—that there are two ways to be wrong with a confidence interval: The interval might be wholly above the mean and miss it by being too far out in the positive tail, or it might be entirely below the mean and miss it by being too far out into the negative side. Since either error is possible, we must utilize both sides of the curve, and this method is called a two-tailed test. In a two-tailed test, the alpha level is split in half and placed in each of the two tails of the distribution, as pictured in Figure 8.1.

Two-tailed tests, then, actually have *two* critical values. The absolute value of these critical values is the same because the curve is symmetrical, but one value is negative and the other is positive. Confidence intervals, by definition, utilize both of the critical values for any alpha level.

Figure 8.1   The Alpha Level and Critical Values of z

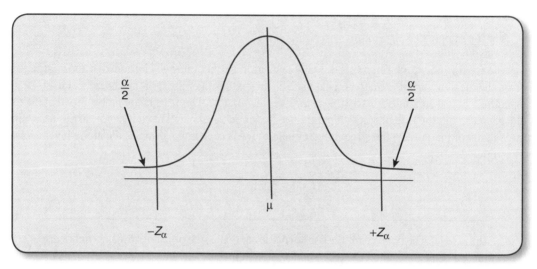

CRITICAL VALUES OF z FOR TWO-TAILED TESTS

CI with 95% confidence $z_\alpha = \pm1.96$

CI with 99% confidence $z_\alpha = \pm2.58$

Now let's consider an example. The Bureau of Justice Statistics maintains a data set called the Census of Jail Inmates (CJI; see Data Sources 8.1). We can use the most recent wave of this census—the year 2005—for an example. This data set is ideal because it contains population data (in other words, it is a census, not a sample, of jails), so we can compute the actual population mean and then pull a random sample, compute its mean and confidence interval, and see if the computed interval contains the true population value.

The variable that will be used is *unconvicted female inmates*, which is the mean number of women being held in pretrial detention across all facilities. The confidence level will be set at 95%.

DATA SOURCES 8.1

THE CENSUS OF JAIL INMATES, 2005

BJS maintains the National Jail Census Series, which is a repository of data sets pertaining to jails across the nation. The CJI is one of the files in this series. The CJI contains inmate data for all local and federal jails, including those operated by private companies, that hold inmates who are either awaiting trial or who have been convicted and are serving a short term of incarceration. Jail-level variables include the total confined population; the confined population broken down by race, gender, and conviction status; and admissions and discharges. All 2,972 jails in operation in June 2005 through March 2006 provided data for the CJI (Bureau of Justice Statistics, 2007); therefore, this is a population rather than a sample.

In the population, the mean number of female inmates who have not been convicted is 18.53 (σ = 60.51). Let us use a sample size of $N = 150$, which is a modest sample size, but is large enough to permit use of the z distribution. The SPSS random sample generator produces a sample with a mean of 28.25 and a standard deviation of 76.24. Since the confidence level is 95%, $z_\alpha = \pm 1.96$. Plugging values for N, $\bar{x}$, s, and z_α into Formula 8(1) yields

$$28.25 \pm 1.96 \left(\frac{76.24}{\sqrt{150-1}} \right) = 28.25 \pm 1.96 \left(\frac{76.24}{12.21} \right)$$

$$= 28.25 \pm 1.96 (6.24) = 28.25 \pm 12.23.$$

The next step is to compute the lower limit (LL) and the upper limit (UL) of the interval, which requires the $\pm$ operation to be carried out, as such:

$$LL = 28.25 - 12.23 = 16.02$$

$$UL = 28.25 + 12.23 = 40.48.$$

Finally, the full interval can be written out:

$$95\% \ CI: 16.02 \le \mu \le 40.48.$$

The interpretation of this interval is that there is a 95% chance (or .95 probability) that the true population mean is 16.02, 40.48, or some number in between those values. Of course, this also means that there is a 5% chance that the true population mean is *not* in this range. In the present example, we know that $\mu = 18.53$, so we can see that here, the confidence interval does indeed contain the population mean. That is good! This sample was one of the 95% that produces accurate confidence intervals rather than one of the 5% that does not. Remember, though, that knowing the value of the true population mean is a luxury that researchers generally do not have; ordinarily, there is no way to check the accuracy of a sample-based confidence interval.

RESEARCH EXAMPLE 8.1

Is there a relationship between unintended pregnancy and intimate partner violence?

Intimate partner violence (IPV) perpetrated by a man against a female intimate is often associated with not just physical violence but a multifaceted web of control in which the woman becomes trapped and isolated. One of the possible consequences is that abused women may not have access to reliable birth control and may therefore be susceptible to unintended pregnancies. These pregnancies, moreover, might worsen the IPV situation because of the emotional and financial burden of pregnancy and childbearing. Martin and Garcia (2011) sought to explore the relationship between IPV and unintended pregnancy in a sample of Latina women in Los Angeles, California. Latinas may be especially vulnerable to both IPV and unintended pregnancy because of the social isolation faced by those who have not assimilated into mainstream U.S. culture. Staff at various prenatal clinics in Los Angeles distributed surveys to their Latina patients. The surveys asked women several questions pertaining to whether or not they intended to get pregnant, whether they experienced emotional or physical abuse by their partner before or during pregnancy, and their level of identification with Mexican versus with Anglo culture. They used a statistic called an *odds ratio*. An odds ratio measures the extent to which consideration of certain independent variables changes the likelihood that the dependent variable will occur. An odds ratio of 1.00 means that the IV does not change the likelihood of the DV. Odds ratios greater than one mean that an IV increases the probability that the DV will occur, and those less than one indicate that an IV reduces the chances of the DV happening.

In Martin and Garcia's study, the DVs in the first analysis were physical and emotional abuse and the DV in the second analysis was physical abuse during pregnancy. The researchers found the following odds ratios and confidence intervals.

	DV: Change in Odds of Unintended Pregnancy	95% CI
IV: Pre-Pregnancy Physical Abuse	.92	.40 ≤ population odds ≤ 2.16
IV: Pre-Pregnancy Emotional Abuse	.50	.26 ≤ population odds ≤ .97

	DV: Change in Odds of Physical Abuse
IV: Unintended Pregnancy	2.80
95% CI	1.01 ≤ population odds ≤ 7.73

Pre-pregnancy physical abuse was not related to the chances that a woman would become pregnant accidentally, as indicated by the odds ratio of .92, which is very close to 1.00. Emotional abuse, surprisingly, significantly reduced the odds of accidental pregnancy (odds ratio = .50). This was an unexpected finding because the researchers predicted that abuse would increase the odds of pregnancy. They theorized that perhaps some emotionally abused women try to get pregnant out of a hope that having a child will improve the domestic situation.

Turning to the second analysis, it can be seen that unintended pregnancy substantially increased women's odds of experiencing physical abuse during pregnancy. Contrary to expectations, women's level of acculturation into either Mexican or Anglo culture did not alter the odds of abuse or pregnancy, once factors such as a woman's age and level of education were accounted for. The findings indicated that the relationship between IPV and unintended pregnancy is complex and deserving of further study in order to identify risk factors for both.

▣ CONFIDENCE INTERVALS FOR MEANS WITH SMALL SAMPLES

The z distribution can be used to construct confidence intervals when $N \geq 100$ because the sampling distribution of means can be safely assumed to be normal in shape. When $N \leq 99$, though, this assumption breaks down and a different distribution is needed. This distribution is the t curve. The *CI* formula for means with small samples is

$$\bar{x} \pm t_\alpha \left(\frac{s}{\sqrt{N-1}} \right), \text{ where}$$

⬅ *Formula 8(3)*

t_α = the critical value of t at a given alpha level.

This formula mirrors that for *CI*s with large samples, with the only difference being that the critical value comes from the t distribution rather than from z.

The critical value of t (t_α) is found using the t table (see Appendix C). There are three pieces of information you need in order to locate t_α. The first is the number of tails. As described above, confidence intervals are always two-tailed, so this is the option that is always used with this type of test. The second determinant of the value of t_α is the alpha level, the computation of which was shown in Formula 8(2). Finally, finding t_α requires you to first compute the degrees of freedom (*df*). With the t distribution, degrees of freedom are related to sample size, as such:

$$df = N - 1.$$

⬅ *Formula 8(4)*

STUDY TIP: THE THREE PIECES OF INFORMATION NEEDED TO READ THE *t* TABLE

We will be using the *t* distribution in later chapters, so it is a good idea to memorize the three pieces of information you always need in order to locate t_α on the *t* table:

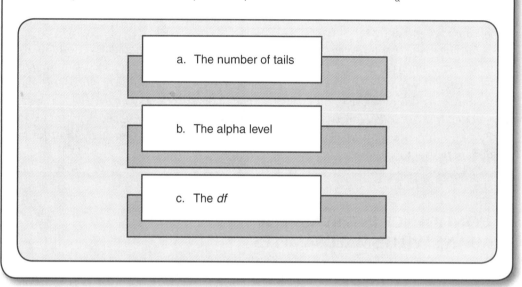

a. The number of tails

b. The alpha level

c. The *df*

The *df* values are located in the rows of the *t* table. Note that not all of the possible values that *df* might assume are included on the table. When the number you are looking for is not there, it is customary to use the largest *df* that is less than the sample-derived *df*. If your sample size was 36, for instance, then your *df* would be 36 − 1 = 35; however, there is no *df* value of 35 on the table, so you would use the *df* = 30 row instead.

For an example of *CIs* with means and small samples, the Firearm Injury Surveillance Study (FISS; see Data Sources 8.2) will be used. This data set contains information about a sample of patients treated in emergency departments (EDs) for gunshot wounds between the years 1993 and 2007. For this example, the sample has been narrowed to those persons who were dead on arrival (DOA) in 2007 as a result of handgun-inflicted wounds. The mean age of this sample ($N = 72$) was 34.10 ($s = 17.00$). The confidence level will be 99%.

DATA SOURCES 8.2

THE FIREARM INJURY SURVEILLANCE STUDY (FISS), 1993–2007

The Centers for Disease Control and Prevention maintain the National Electronic Injury Surveillance System, of which the FISS is a part. The data are from a nationally

(Continued)

representative sample of hospitals stratified by size. Data include only those patients who did not die in the emergency department (ED) in which they were treated; those who died prior to arrival are assigned DOA status, while those who died after being transferred from the ED to some other hospital unit are coded as transfers (Centers for Disease Control and Prevention, 2010). This is a limitation of the FISS that circumscribes its usefulness in criminal justice research, but this data set is nonetheless valuable as one of the few major studies that systematically tracks gun-related injuries.

The first thing that must be done to construct the *CI* is to find t_α. Going to the *t* table and using the three necessary pieces of information (this is a two-tailed test with $\alpha = 1 - .99 = .01$ and $df = 72 - 1 = 71$), you can see that $df = 71$ is not located on the table, so we must use the $df = 60$ row instead. The critical value of *t* is ± 2.660.

Next, plug all the numbers into the formula:

$$34.10 \pm 2.660\left(\frac{17.00}{\sqrt{72-1}}\right) = 34.10 \pm 2.660\left(\frac{17.00}{8.43}\right) = 34.10 \pm 2.660(2.02) = 34.10 \pm 5.37.$$

Compute the lower and upper limits:

$$LL = 34.10 - 5.37 = 28.73$$
$$UL = 34.10 + 5.37 = 39.47.$$

Finally, assemble the full confidence interval:

$$99\%\ CI: 28.73 \leq \mu \leq 39.47.$$

There is a 99% chance that the mean age of handgun injury victims who arrived at the emergency room DOA in 2007 was between 28.73 and 39.47, inclusive.

You may have noted that this is not a very good estimate—there are nearly 11 years included in this range of possible values of μ! This *CI* lacks precision. The likely reason for this is the small sample size; with only 72 cases available for analysis, it is difficult to pin down a good estimate of the population mean. Precision also suffered when the relatively demanding 99% confidence standard was selected.

RESEARCH EXAMPLE 8.2

Why Do Suspects Confess to Police?

The U.S. Constitution's Fifth Amendment protection from compelled self-incrimination ensures that persons suspected of having committed criminal offenses do not have to

(Continued)

speak to the police or answer any questions the police may ask. Despite this right, a good number of suspects do talk to police and do provide incriminating statements. As the wry saying goes, "Everyone has the *right* to remain silent, but not everyone has the *ability* to do so." So what makes suspects confess? Deslauriers-Varin, Beauregard, and Wong (2011) sought to identify some of the contextual factors that make suspects more likely to confess, even when those suspects initially indicated that they wished to remain silent. The researchers obtained a sample of 211 convicted male offenders from a Canadian prison and gathered extensive information about each participant. The researchers analyzed odds ratios, just as the study in Research Example 8.1 did. Recall that an odds ratio of 1.00 means that the independent variable does not alter the odds that the dependent variable will occur. Odds ratios less than 1.00 mean the IV makes the DV less likely, whereas odds ratios greater than 1.00 indicate that the IV makes the DV more likely to happen. The researchers found the following odds ratios and confidence intervals.

| | DV: Suspect Did Not Confess | |
Independent Variable	*Change in Odds of No Confession*	*95% CI*
Initial Decision: No Confession	25.33*	10.05, 63.82
Criminal History		
1 or 2 Prior Convictions	3.01	.79, 11.42
3+ Prior Convictions	13.29*	3.05, 57.91
Had an Accomplice	2.58*	1.02, 6.52
Police Evidence Is Strong	.23*	.10, .55
Lawyer Advised Nonconfession	3.29*	1.34, 8.07
Crime Was Drug-Related	3.55*	1.26, 10.02

Source: Adapted From Table 2 in Deslauriers-Varin, Beauregard, and Wong (2011).

The numbers in the above table that are flagged with asterisks are those that are statistically significant, meaning that the IV exerted a noteworthy impact on suspects' decision to remain silent rather than confessing. The initial decision to not confess was the strongest predictor of suspects' ultimate refusal to provide a confession; those who initially resisted confessing were likely to stick to that decision. Criminal history was also related—having only one or two priors was not related to nonconfession, but suspects with three or more prior convictions were substantially more likely to remain silent. This may be because these suspects were concerned about being sentenced harshly as habitual offenders. The presence of an accomplice also made nonconfession more likely, as did a lawyer's advice to not confess. Those accused of drug-related crimes were more likely to not confess, though crime type was not significant for suspects accused of other types of offenses. Finally, the strength of police evidence was a factor in suspects' decisions. Strong police evidence was related to a significant reduction in the odds of nonconfession (that is, strong evidence resulted in a greater chance that the suspect would confess).

It appeared, then, that there are many factors that impact suspects' choice regarding confession. Most of these factors appear to be out of the control of police interrogators; however, the researchers did not include variables measuring police behavior during interrogation, so there may well be techniques police can use to elicit confessions even from those suspects who are disinclined to offer information. One policy implication from these results involves the importance of the initial decision in the final decision—79% of offenders stuck with their initial decision concerning whether or not to confess. Police might, therefore, benefit from focusing their efforts on influencing the initial decision rather than allowing a suspect to formulate a decision first and then applying interrogation tactics.

◙ CONFIDENCE INTERVALS WITH PROPORTIONS AND PERCENTAGES

The principles that guide the construction of confidence intervals around means also apply to confidence intervals around proportions and percentages. There is no difference in the procedure used for proportions versus that for percentages—percentages simply have to be converted to proportions before they are plugged into the *CI* formula. For this reason, we will speak in terms of proportions for the remainder of the discussion.

There are sampling distributions for proportions just as there are for means. Because of sampling error, a sample proportion (symbolized $\hat{p}$; pronounced "*p* hat") cannot be assumed to equal the population proportion (P), so confidence intervals must be created in order to estimate the population values with a certain level of probability.

Confidence intervals for proportions employ the z distribution. The normality of the sampling distribution of sample proportions is a bit iffy, but generally speaking, z is a safe bet as long as the sample is large and contains at least five successes and at least five failures. The formula for confidence intervals with proportions is

$$\hat{p} \pm z_\alpha \sqrt{\frac{\hat{p}(1-\hat{p})}{N}}, \text{ where} \qquad \boxed{Formula\ 8(5)}$$

$\hat{p}$ = the sample proportion.

To illustrate *CIs* for proportions with large samples, we will again use the 2007 FISS. This time, we will examine the involvement of handguns in all nonfatal firearm injuries (though this just means that the patients did not die in the ED; they may have died prior to arriving or after leaving). According to the FISS, handguns were the mechanism of injury in 56.45% of all firearm-related wounds for which gun type was known (this category includes BB guns). The sample size is $N = 2,000$. Given this large sample size and the fact that the number of successes (defined here as the mechanism of injury being a handgun) and failures (here, the mechanism being a firearm of some other type) both well exceed the minimum of five, the z distribution can be used and the analysis can proceed. Confidence will be set at 99%, which means $z_\alpha = \pm 2.58$.

First, the sample percent needs to be converted to a proportion, like so: $\dfrac{56.45}{100} = .56$. Plugging the numbers into Formula 8(5) and solving:

$$.56 \pm 2.58 \sqrt{\frac{.56(1-.56)}{2000}} = .56 \pm 2.58 \sqrt{\frac{.56(.44)}{2000}} = .56 \pm 2.58 \sqrt{\frac{.25}{2000}}$$

$$= .56 \pm 2.58 \sqrt{.0001} = .56 \pm 2.58(.01) = .56 \pm .03$$

$$LL = .56 - .03 = .53$$

$$UL = .56 + .03 = .59$$

$$99\% \ CI: .53 \leq P \leq .59.$$

We can say with 99% certainty that in 2007, handguns accounted for between .53 and .59, inclusive, of firearm-related injuries, where firearm type was known.

CHAPTER SUMMARY

Confidence intervals are a way for researchers to use sample statistics to form conclusions about the probable values of population parameters. Confidence intervals entail the construction of ranges of values predicted to contain the true population parameter. The researcher sets the level of confidence (probability of correctness) according to her or his judgment about the relative costs of a loss of confidence versus compromised precision. The conventional confidence levels in criminal justice/criminology research are 95% and 99%. The decision about level of confidence must be made with consideration to the tradeoff between confidence and precision—as confidence increases, the quality of the estimate diminishes. All confidence intervals are two-tailed, which means that alpha is divided in half and placed in both tails of the distribution. This creates two critical values. The critical values have the same absolute value, but one is negative and one is positive.

There are two types of CIs for means: large sample and small sample. Confidence intervals for means with large samples employ the z distribution, while those for means with small samples use the t curve. When the t distribution is used, it is necessary to calculate degrees of freedom in order to locate the critical value on the t table.

Confidence intervals can be constructed on the basis of proportions, providing that two criteria are met. First, the sample must contain at least 100 cases. Second, there must be at least five successes and five failures in the sample. These two conditions help ensure the normality of the sampling distribution and, thus, the applicability of the z curve.

CHAPTER 8 REVIEW PROBLEMS

Answer the following questions with regard to confidence intervals.

1. How many cases must be in a sample for that sample to be considered "large"?

2. "Small" samples are those that have ____ or fewer cases.

3. Which distribution is used with large samples?

4. Which distribution is used with small samples?

5. Why can the distribution that is used with large samples not also be used with small ones?

6. Explain the tradeoff between confidence and precision.

7. An item in the Police–Public Contact Survey (PPCS) asked respondents ($N = 11,843$) who had experienced a recent face-to-face encounter with police to report the number of encounters they had in the past 6 months. The mean number of contacts was 1.60 with a standard deviation of 2.40. Construct a 95% confidence interval around this sample value, and interpret that interval in words.

8. PPCS respondents were surveyed about whether police had used force against them during a traffic stop in the past 12 months and whether they believed that force to have been excessive. The 28 respondents who thought the force was excessive reported a mean of 2.18 officers on the scene of the incident ($s = 1.47$). Construct a 95% confidence interval around this sample value, and interpret the interval in words.

9. Of the PPCS respondents who were arrested during the course of a traffic stop ($N = 114$), 77 reported that the officer on the scene was respectful during the encounter. Construct a 99% confidence interval around this sample value, and interpret the interval in words. (Note: You first must find $\hat{p}$.)

10. In the 2004 National Crime Victimization Survey (NCVS), respondents reported a mean of 1.59 criminal victimizations within the past 6 months ($s = 1.18$; $N = 15,142$). Construct a 99% confidence interval around this sample value, and interpret that interval in words.

11. In the 2004 NCVS, 60 percent of respondents who had been the victim of a personal attack ($N = 1,262$) said that they did not suffer any injuries as a result of the assault. Construct a 95% confidence interval around this sample value, and interpret the interval in words.

12. According to the Firearm Injury Surveillance Study's 2007 data, 1,055 people in the sample were treated in emergency rooms for wounds inflicted during criminal assaults with firearms. In 147 of these assault cases, the assailant was a friend, acquaintance, or family member of the victim. Construct a 95% confidence interval around this sample value, and interpret the interval in words.

13. In a random sample of 150 jails from the Census of Jail Inmates, 2005, 53% of inmates housed in these facilities were being held in pretrial detention (had not been convicted of a crime). Construct a 99% confidence interval around this sample value, and interpret the interval in words.

14. According to the 2003 Law Enforcement Management and Administrative Statistics (LEMAS) survey, there was a mean of 1.32 ($s = .85$) officers per 1,000 citizens across the 201 cities with populations between 150,000 and 300,000. Construct a 95% confidence interval around this sample value, and interpret the interval in words.

15. According to the 2003 LEMAS survey, cities with populations of 1 million or more ($N = 84$) had .70 ($s = .96$) officers per 1,000 citizens. Construct a 99% confidence interval around this sample value, and interpret the interval in words.

One question of substantial interest to criminology and to public policy is the potential for racial differences in pretrial detention. Defendants who are detained prior to trial are more likely to be convicted because they are unable to effectively participate in their own defense. Those detained also face possible job loss, family disruption, and physical danger. Racial differences in pretrial detention, then, could have resounding consequences. The State Court Processing Statistics data set contains information on defendant race/ethnicity and pretrial detention. The sample has been narrowed to male defendants facing charges for violent felonies. Construct the following three confidence intervals.

16. Among white males accused of violent felonies ($N = 806$), .42 were detained prior to trial. Construct a 95% confidence interval, and interpret this interval in words.

17. Among black males accused of violent felonies ($N = 1,443$), .53 were detained prior to trial. Construct a 95% confidence interval, and interpret this interval in words.

18. Among Hispanic males accused of violent felonies ($N = 836$), .50 were detained prior to trial. Construct a 95% confidence interval, and interpret this interval in words.

In the 2004 General Social Survey, respondents were asked how they feel toward Muslims on a 0 to 100 "temperature" scale with 0 = very cold; 100 = very warm. They were also asked about their feelings toward capital punishment (coded as favor/oppose). Construct the following three confidence intervals.

19. The respondents who favored the death penalty for people convicted of murder ($N = 511$) had a mean of 46.14 ($s = 22.69$) on the scale measuring attitudes toward Muslims. Construct a 99% confidence interval around this sample value, and interpret the interval in words.

20. Those opposing the death penalty ($N = 240$) had a mean of 50.24 ($s = 22.68$) on the scale measuring attitudes toward Muslims. Construct a 99% confidence interval around this sample value, and interpret the interval in words.

21. The respondents who were unsure about their attitudes toward capital punishment ($N = 35$) had a mean of 55.00 ($s = 19.55$) on the scale. Construct a 99% confidence interval around this sample value, and interpret the interval in words.

KEY TERMS

Sampling error	Level of confidence	Two-tailed test
Point estimate	Alpha level	
Confidence interval	Critical value	

GLOSSARY OF SYMBOLS INTRODUCED IN THIS CHAPTER

α The probability that a *CI* does not contain the population parameter

z_α The *z* score associated with a given α level

t_α The *t* score associated with a given α level

$\hat{p}$ A sample proportion

P A population proportion

REFERENCES

Bureau of Justice Statistics. (2007). *Census of jail inmates: Individual-level data, 2005* [computer file]. Ann Arbor, MI: Inter-University Consortium for Political and Social Research. DOI:10.3886/ICPSR20367.v1

Centers for Disease Control and Prevention. National Center for Injury Prevention and Control. (2010). *Firearm injury surveillance study, 1993–2007* [computer file]. Ann Arbor, MI: Inter-University Consortium for Political and Social Research. DOI:10.3886/ICPSR27002.v1

Deslauriers-Varin, N., Beauregard, E., & Wong, J. (2011). Changing their mind about confessing to police: The role of contextual factors in crime confession. *Police Quarterly, 14(5)*, 5–24.

Martin, K. R., & Garcia, L. (2011). Unintended pregnancy and intimate partner violence before and during pregnancy among Latina women in Los Angeles, California. *Journal of Interpersonal Violence, 26*(6), 1157–1175.

PART III

HYPOTHESIS TESTING

Y ou have now learned about descriptive statistics (Part I) and the theories of probability and distributions that form the foundation of statistics (Part II). Part III brings all of this together to form what most criminal justice/criminology researchers consider to be the high point of statistics: **inferential analyses** or what is also frequently called hypothesis testing. Hypothesis testing involves the use of a sample as a means of drawing a conclusion about a population. The basic strategy is a two-step process. First, a sample statistic is computed. Second, a probability distribution is used to find out whether this sample statistic has a low or high probability of occurrence. It is this probability assessment that guides researchers in making decisions and reaching conclusions. Samples are thus vehicles that allow you to make generalizations about what you believe is happening in the population as a whole.

Inferential analysis: The process of generalizing from a sample to a population; the use of a sample statistic to estimate a population parameter. Also called hypothesis testing in reference to the process of testing the validity of a theoretical concept using a sample and inferential procedures.

Part III covers some of the bivariate (involving two variables) inferential tests commonly used in criminology/criminal justice research: chi-square tests of independence, two-population tests for differences between means and proportions, analyses of variance, and correlations. The book ends with an introduction to bivariate and multiple regression.

The proper test to use in a given hypothesis-testing situation is determined by the level of measurement of the variables with which you are working. If your memory of levels of measurement has become a bit fuzzy, go back now and review this important topic. You will not be able to select the correct analysis unless you can identify your variables' levels of measurement. Figure 9.1 shows the levels of measurement. For purposes of hypothesis testing, the key distinction is that between categorical and continuous variables. Be sure you can accurately identify variables' level of measurement before you start Chapter 10.

Levels of Measurement

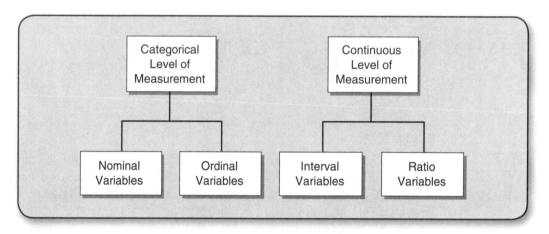

STUDY TIP: REPETITION IS KEY

There is no sense in denying that you may find many of the concepts presented in the following chapters foreign and confusing at first. Take heart! The process of learning statistics centers on repetition. Read and reread the chapters, study your lecture notes, and do the end-of-chapter review problems; things will start to sink in. Terms, formulas, and ideas that initially seemed incomprehensible will gradually take form in your mind and begin making sense. Remember that every criminology/criminal justice researcher started off in a position much like yours! There was a point when they knew nothing about statistics and had to study hard to construct a knowledge base. Commit the time and effort and there is a good chance that you will be pleasantly surprised by how well you will do.

Chapter 9

Hypothesis Testing

A Conceptual Introduction

T he purpose of this chapter is to provide you with a clear conceptual foundation of the nature and purpose of hypothesis testing. It is worth developing a solid understanding of the substance of inferential statistics before approaching the specific types of hypothesis tests covered in the proceeding chapters. This chapter will help you grasp the overarching logic behind these sorts of tests so that you can approach the "trees" (specific tests) with a clear picture of what the "forest" (underlying conceptual framework) looks like.

By now, you should be very familiar with the idea that researchers are usually interested in populations but, because populations are so large, samples must suffice instead. It is not typically the samples themselves that are of interest, though—the ultimate goal for the quantitative researcher is to generalize from the sample to the population. Hypothesis testing is the process of making this generalization.

What we are really talking about in inferential statistics is the probability of empirical outcomes. There are infinite samples that can be drawn from any given population and, therefore, there are infinite sample statistics that could be drawn as well. When we have a sample statistic that we wish to use inferentially, the question asked is, "Out of all the samples and sample statistics possible, what is the probability that I would draw *this* one?"

**STUDY TIP: EXPECTED VERSUS OBSERVED/
EMPIRICAL OUTCOMES**

Remember the difference between expected and observed or empirical outcomes. Expected outcomes are the results you anticipate seeing on the basis of probability theory. You constructed a table of expected outcomes in the context of binomials. Observed outcomes, by contrast, are what you *actually* see. These results may or may not match expectations.

Here is another way to think about it: You know that observed outcomes often differ from what is expected on the basis of theory, prior research, or other knowledge sources. Any time a sample statistic is not equal to a population parameter, there are two potential explanations for the difference. First, it may be the product of the random fluctuations in sample statistics that is called sampling error. In other words, the disparity may simply be a meaningless fluke. The second possible explanation for the difference is that there actually is a genuine discrepancy between the sample statistic and the population parameter. In other words, the disparity could represent a bona fide statistical effect.

When researchers first approach an empirical finding, they do not know which of the above two explanations accounts for the observed or empirical result. The overarching purpose of hypothesis testing is to determine which of them appears to be the more valid of the two. This is where probabilities come in. Researchers identify the probability of observing a particular empirical result and then use that probability to make a decision about which explanation seems to be correct.

This is pretty abstract, so an example is in order. We will use the BJS Census of State and Federal Adult Correctional Facilities (CSFACF; Data Sources 7). The CSFACF contains information about the rate of inmate-on-inmate assaults at each facility and whether or not that facility is under court order for inadequate staffing. We might hypothesize that because staffing seems clearly related to internal institutional security, those facilities under court order for severely deficient staffing levels would have higher inmate-on-inmate assault rates than the facilities that are not under staffing orders. In other words, we want to know if there is a relationship between critically deficient correctional staffing (the independent variable, or IV) and inmate-on-inmate violence (the dependent variable, or DV).

First, we need the mean assault rate for each facility classification. Those institutions that were not under court order experienced a mean of 1.75 prisoner assaults per 100 prisoners housed in the facility during the study year, while those that were under court order reported a mean of 2.02 assaults per 100. These are the empirical findings, and you can see that they are unequal; however, recall that there are *two potential reasons* for this disparity. Their inequality might be meaningless and the differences between the numbers purely the product of chance; in other words, the finding might be a fluke. On the other hand, they might be unequal because staffing really is related to prisoner assaults and there is, therefore, a significant difference between the means. These two competing potential explanations can be framed as hypotheses.

The first of these hypotheses is called the **null hypothesis**. The null (symbolized H_0) represents the prediction that the difference between the population and sample is purely the product of sampling error or, in other words, chance variation in the data. You can use the word *null* as its own mnemonic device because this word means "nothing." Something that is null is devoid of meaning. In the context of the present example, the null predicts that staffing is not related to inmate-on-inmate assaults and that the observed difference between the means is just chance error.

The second possible explanation is that prisons under court order for staffing really do have a higher mean assault rate. This prediction is spelled out in the **alternative hypothesis** (symbolized H_1). The alternative is, essentially, the opposite of the null: The null predicts that there is no relationship between the two variables being examined, and the alternative predicts that they are related.

Null hypothesis: In an inferential test, the hypothesis predicting that there is no relationship between the independent and dependent variables. Symbolized H_0.

Alternative hypothesis: In an inferential test, the hypothesis predicting that there is a relationship between the independent and dependent variables. Symbolized H_1.

In the context of the present example, the null and alternative can be written as:

H_0: Prisons under court order for staffing-related problems have the same mean assault rate as those prisons not under court order; that is, the two means are equal and there is no staffing-assault relationship.

H_1: Prisons under court order for staffing-related problems have a higher mean assault rate than do those prisons not under court order; that is, there is a relationship between staffing and assaults, with understaffed facilities experiencing significantly more assaults.

More common than writing them out in words is to use symbols to represent the ideas embodied in the above longhand versions. Transforming these concepts into such symbols turns the null and alternative into:

$$H_0: \mu_1 = \mu_2$$
$$H_1: \mu_1 > \mu_2, \text{ where}$$

μ_1 = the mean number of assaults in prisons that were under court order for staffing,

μ_2 = the mean number of assaults in prisons that were not under court order for staffing.

If the null is true and there is, in fact, no staffing-assault relationship, then we would conclude that all of the prisons come from the same population. If, instead, the alternative is true, then there are actually two populations at play here—one that is under court order for staffing and one that is not. Figure 9.1 shows this idea pictorially.

Figure 9.1 One Population or Two?

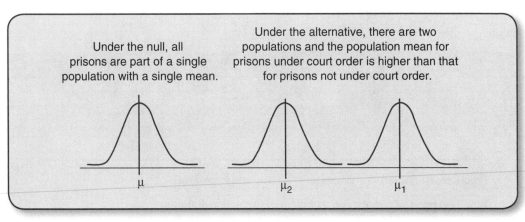

The assumption going into an inferential analysis is that the null is the true state of affairs. In other words, the default assumption is that there is no relationship between the independent and dependent variables. The overarching goal in conducting the test is to decide whether to retain the null (concluding that there is indeed no IV-DV relationship) or to reject the null (concluding that there is, in fact, a relationship between the variables). The null can only be rejected if there is solid, compelling evidence that leads you to decide that this hypothesis is simply not accurate.

A good analogy to the logic behind hypothesis testing is the presumption of innocence in a criminal trial. At the outset of a trial, the jury must consider the defendant to be legally innocent of the crime of which she or he is accused. The "null" here is innocence and the "alternative" is guilt. If the prosecutor fails to convincingly show guilt, then the innocence assumption stands and the defendant must be acquitted. If, however, the prosecutor presents sufficient incriminating evidence, then the jury rejects the assumption of innocence and renders a guilty verdict.

In inferential statistics, researchers approach a hypothesis test assuming that the null is true. They construct a probability framework based upon that assumption. The question they are trying to answer is, "What is the probability of observing the empirical result that I see in front of me *if* the null hypothesis is true?" If the probability of the null being true is extremely low, then the null is rejected because it strains the imagination to think that something with such a low likelihood of being true actually is the correct explanation for an empirical phenomenon. The alternative would, thus, be taken as being the more likely version of reality. If the probability of the null being true is not low, then the null is considered to be a viable explanation for the results and it is retained.

Of course, there is always a chance that a researcher's decision regarding whether to reject or retain the null is wrong. There are two types of errors that can be made in this regard. A **Type I** error occurs when a true null is erroneously rejected, while a **Type II** error happens when a false null is inaccurately retained. A Type I error corresponds to a wrongful conviction of an innocent defendant and a Type II error is analogous to a wrongful acquittal of a guilty defendant. Researchers can often minimize the probability that they are wrong about a decision, but they can never eliminate it. For this reason, you should always be circumspect as both a producer and a consumer of statistical information. You should never rush haphazardly to conclusions. Anytime you or anyone else runs a statistical analysis and makes a decision about the null hypothesis, there is a probability—however minute it may be—that that decision is wrong.

Table 9.1   Type I and Type II Errors

	... the null is actually false	*... the null is actually true*
If you reject the null and ...	Correct!	Type I Error
If you retain the null and ...	Type II Error	Correct!

Type I error: The erroneous rejection of a true null hypothesis.

Type II error: The erroneous retention of a false null hypothesis.

This book will cover several different types of hypothesis testing procedures in the bivariate context. The choice between them is made upon the basis of each variable's level of measurement. You must identify the levels of measurement of the independent and dependent variables and then select the proper test for those measurement types. Table 9.2 is a handy chart which you should study closely and refer to repeatedly throughout the next few chapters.

Table 9.2 Choosing the Appropriate Bivariate Test Based on the Variables' Level of Measurement

	The Independent Variable is . . .	
The Dependent Variable is . . .	*Categorical*	*Continuous*
Categorical	Chi-square	*t* test ANOVA
Continuous	N/A	Correlation

If you choose the wrong test, you will arrive at an incorrect answer. Period, no two ways around it; the result will be wrong. This is true in both hand calculations and SPSS programming. SPSS rarely gives error messages and will usually run analyses even when they are deeply flawed. Remember, GIGO! When garbage is entered into an analysis, the output is garbage as well. You must be knowledgeable about the proper use of these statistical techniques, or you risk becoming either a purveyor or consumer of erroneous results.

The final topic of this chapter is an introduction to the steps of hypothesis testing. It is useful to outline a framework by which you can abide consistently as you learn the different types of analyses. This lends structure to the learning process and allows you to see the similarities between various techniques. Hypothesis testing is broken down into five steps, as follows.

Step 1. State the null (H_0) and alternative (H_1) hypotheses.

- The two competing hypotheses that will be tested are laid out.

Step 2. Identify the distribution and compute the degrees of freedom.

- Each type of statistical analysis utilizes a certain probability distribution. You have already seen the *z* and *t* distributions, and more will be introduced in later chapters. You have encountered the concept of degrees of freedom (*df*) in the context of the *t* distribution. Other distributions also require the computation of *df*.

Step 3. Identify the critical value of the test statistic and state the decision rule.

- The critical value is based upon probability. The critical value is the number that the obtained value (the result of the analytical portion of the test conducted in Step 4) must exceed in order for the null to be rejected. The decision rule is an a priori statement formally laying out the criteria that must be met for the null to be rejected.

Step 4. Compute the obtained value of the test statistic.

- This is the analytical heart of the hypothesis test. You will select the appropriate formula, plug in the relevant numbers, and solve. The outcome is the obtained value of the test statistic.

Step 5. Make a decision about the null and state the substantive conclusion.

- You will revisit your decision rule from Step 3 and decide whether to reject or retain the null based on the comparison between the critical and obtained values of the test statistic. Then you will render a substantive conclusion. Researchers have the responsibility to interpret their statistical findings and draw substantive conclusions that make sense to other researchers and to the general public.

CHAPTER SUMMARY

This chapter provided an overview of the nature, purpose, and logic of hypothesis testing. The goal of statistics in criminology/criminal justice is usually generalization from a sample to a population. This is accomplished by first finding a sample statistic and then determining the probability that that statistic would be observed by chance alone. If the probability of the result being attributable solely to chance is exceedingly low, then the researcher concludes that the finding is not due to chance and is, instead, a genuine effect.

When a researcher has identified an independent and dependent variable that she or he thinks might be related, there are two possibilities with respect to that IV and DV. The first possibility is that they are, in fact, not related. This possibility is embodied by the null hypothesis, symbolized H_0. The second possibility is that the IV and DV are related. This is the alternative hypothesis, H_1. A hypothesis test using the five steps outlined in this chapter will ultimately result in the null being either rejected or retained and the researcher concluding that the IV either are or are not, respectively, related to one another.

CHAPTER 9 REVIEW PROBLEMS

1. Suppose a researcher was studying gender differences in sentencing. She found that males sentenced to jail received a mean of 6.45 months and females sentenced to jail received a mean of 5.82 months. Describe the two possible reasons for the differences between these two means.

2. Write the symbol for the null hypothesis and explain what this hypothesis predicts.

3. Write the symbol for the alternative hypothesis and explain what this hypothesis predicts.

4. How does the null hypothesis relate to the presumption of innocence in a criminal trial?

5. Explain what a Type I error is.

6. Explain what a Type II error is.

7. Define the word *bivariate*.

8. If you computed an empirical result, identified the probability of observing that result, and found that the probability was high, would you conclude that this is an atypical, unusual result, or would this be a not-unexpected finding? Would you conclude that this is the product of sampling error, or would you think that it is a true effect?

9. If you computed an empirical result, identified the probability of observing that result, and found that the probability was very low, would you conclude that this is an atypical, unusual result, or would this be an unsurprising or not-expected finding? Would you conclude that this is the product of sampling error, or would you think that it is a true effect?

10. Identify the correct hypothesis-testing procedure or procedures for a categorical IV and a continuous DV.

11. Identify the correct hypothesis-testing procedure for a categorical IV and a categorical DV.

12. Identify the correct hypothesis-testing procedure for a continuous IV and a continuous DV.

13. Suppose a researcher is testing for a relationship between race and sentencing. She measures race as *black; Hispanic/Latino; white; other* and sentencing as *prison; jail; fine; other*. Based on Table 9.2 and these variables' levels of measurement, what would be the correct hypothesis-testing procedure?

14. Suppose a researcher is testing for a relationship between education level and criminal offending. He measures education as *number of years of schooling completed* and offending as *number of crimes committed*. Based on Table 9.2 and these variables' levels of measurement, what would be the correct hypothesis-testing procedure?

KEY TERMS

Inferential analysis	Alternative hypothesis	Type II error
Null hypothesis	Type I error	

GLOSSARY OF SYMBOLS INTRODUCED IN THIS CHAPTER

H_0	The null hypothesis
H_1	The alternative hypothesis

Chapter 10

Hypothesis Testing With Two Categorical Variables

Chi-Square

T he first type of bivariate hypothesis test presented is the **chi-square test of independence**. The distinguishing feature of this test is that both the independent variable (IV) and the dependent variable (DV) are categorical. Before going into the theory and math behind the chi-square statistic, read the Research Examples for illustrations of the types of situations in which a criminal justice/criminology researcher would utilize a chi-square test.

Chi-square test of independence: The hypothesis testing procedure appropriate when the independent and dependent variables are both categorical.

RESEARCH EXAMPLE 10.1

Do Traffic Stops Alter People's Likelihood of Calling the Police?

Police rely on local community residents to call in when they see problematic people or conditions in their neighborhoods. Effective crime control and order maintenance depends, in part, on people's willingness to voluntarily make these types of reports.

There are many reasons, however, why people may be loath to call the police to report problems. Gibson, Walker, Jennings, and Miller (2010) hypothesized that one factor that could deter people from reporting neighborhood problems is having had a recent negative encounter with a police officer. Using the 1999 wave of the Police–Public Contact Survey, the researchers identified three variables. The primary independent variable was, "How many times in the past 12 months were you in a vehicle that was stopped by police?" Respondents' answers were coded as *none*, *once*, and *more than once*. Two dependent variables were used to tap into citizens' willingness to alert police to the presence of local issues. The first was, "In the past 12 months, did you contact the police to ask for assistance or information?" and the second was "In the past 12 months, did you contact the police to report a neighborhood problem?" Both variables were coded as *not at all* and *one or more times*. The following are bivariate contingency tables (a.k.a. crosstabs) displaying the overlap between the predictor and each of the outcome measures.[1] How can the researchers find out if the two variables are related?

Tables 10.1a and 10.1b   Contingency Tables for Having Experienced a Traffic Stop and Having Called the Police for Assistance or to Report Neighborhood Problems

	Number of Times Called for Assistance		
Number of Stops	*None*	*One or More*	*Row Marginal*
None	6,068	1,433	7,501
One	5,639	404	6,043
More than One	1,289	136	1,425
Column Marginal	12,996	1,973	$N = 14,969$
	Number of Times Called to Report Problems		
Number of Stops	*None*	*One or More*	*Row Marginal*
None	6,409	1,095	7,504
One	5,724	323	6,047
More than One	1,335	92	1,427
Column Marginal	13,468	1,510	$N = 14,978$

RESEARCH EXAMPLE 10.2

Do Waived Juvenile Defendants' Race/Ethnicity Affect the Sentences These Juveniles Receive?

Several criminology/criminal justice studies have demonstrated that black and Hispanic/Latino defendants are sentenced more harshly than their white counterparts, even when factors such as the severity of the instant offense and defendants' prior criminal records are accounted for. There has not, however, been attention paid to the effects of race and ethnicity on sentencing for juvenile defendants who are waived into criminal courts to be tried as adults. Jordan and Freiburger (2010) wished to address this gap in the literature by studying the sentences received by waived juveniles of multiple races and ethnicities. The authors selected the 1998 Juvenile Defendants in Criminal Courts data set (see Data Sources 10 later in this chapter), which offers information about juveniles in 40 urban counties who were charged with felonies and were either transferred or waived to adult court. The researchers' independent variable was *race and ethnicity* and was measured as *white*, *black*, or *Hispanic.* The dependent variable was *sentence received* and was coded as *probation*, *jail*, or *prison*. (There were so few juveniles of other races and ethnicities that the authors decided to exclude them from the analysis.) The following crosstabs table displays these two variables' joint distribution.[2] How can the researchers find out of there is a relationship between the two variables?

Table 10.2   Contingency Table for Juvenile Race or Ethnicity and Sentence Received

Race or Ethnicity	Sentence Received			
	Jail	Prison	Probation	Row Marginal
White	226	102	333	661
Black	398	295	917	1,610
Hispanic	177	83	195	455
Column Marginal	801	480	1,445	$N = 2,726$

In the study summarized in Research Example 10.1, *number of stops* is the independent variable and is ordinal. *Number of times called for assistance* and *number of times called to report neighborhood problems* are the dependent variables and are also ordinal. In the second study, *race or ethnicity* (the IV) and *sentence received* (the DV) are both nominal. The question each study attempted to answer was, "Are these two variables related? That is, does the independent variable appear to exert an impact on the dependent variable?" Answering this question requires the use of the chi-square test of independence because in both of these studies, the IV and DV (or DVs) are categorical.

▣ CONCEPTUAL BASIS OF THE CHI-SQUARE TEST: STATISTICAL DEPENDENCE AND INDEPENDENCE

Two variables that are not related to one another are said to be **statistically independent**. When two variables are related, they are said to be **statistically dependent**. Statistical independence means that knowing which category a person or object falls into on the IV does not help predict their placement on the DV. When, conversely, two variables are statistically dependent, the IV does have predictive power over the DV. In Research Example 10.1, the independent variable was the number of traffic stops survey respondents reported having experienced in the past 12 months, and one of the the dependent variables was whether or not respondents had called the police to request assistance. If these two variables are statistically independent, then having experienced a traffic stop will *not* influence respondents' likelihood of calling the police; if the variables are statistically dependent, then knowing whether or not someone has been the subject of a traffic stop will help predict whether that person has called the police for assistance or to report problems. When we want to know whether there is a relationship between two categorical variables, we turn to the chi-square test of independence.

Statistical independence: The condition in which two variables are not related to one another; that is, knowing what classpersons or objects fall into on the IV does not help predict which class they will fall into on the DV.

Statistical dependence: The condition in which two variables are related to one another; that is, knowing what class persons or objects fall into on the IV helps predict which class they will fall into on the DV.

▣ THE CHI-SQUARE TEST OF INDEPENDENCE

Let us work slowly through an example of a chi-square hypothesis test using the five steps described in Chapter 9 and discuss each step in detail along the way. For this example, we will turn to the 2006 General Social Survey (GSS; see Data Sources 2.2) and the issue of gender differences in attitudes about crime and punishment. Theory is somewhat conflicting as to whether women tend to be more forgiving of transgressions and to prefer leniency in punishment or whether they generally prefer harsher penalties out of the belief that offenders pose a threat to community safety. The GSS contains data on the sex of respondents and these persons' attitudes toward the death penalty. The joint frequency distribution is displayed in Table 10.4. We will test for a relationship between gender and death penalty support using the chi-square test of independence.

Step 1. State the Null (H₀) and Alternative (H₁) Hypotheses

The null hypothesis (H_0) in chi-square tests is that there is no relationship between the independent and dependent variables. The chi-square test statistic is χ^2 (χ is the Greek letter *chi* and is pronounced "kye"). A χ^2 value of zero means that the variables are unrelated, so the null is formally written as

$$H_0: \chi^2 = 0.$$

Table 10.3   2006 GSS Respondents' Sex and Preference for the Death Penalty

| Sex | Attitude Toward Death Penalty for Persons Convicted of Murder | | Row Marginal |
	Favor	Oppose	
Female	1,010[A]	573[B]	1,583
Male	875[C]	357[D]	1,232
Column Marginal	1,885	930	$N = 2,815$

The alternative hypothesis (H_1), on the other hand, predicts that there is a relationship. The chi-square statistic gets larger as the overlap or relationship between the IV and DV increases. The alternative is

$$H_1: \chi^2 > 0.$$

Step 2. Identify the Distribution and Compute the Degrees of Freedom

The χ^2 statistic has its own theoretical probability distribution—the χ^2 distribution. The χ^2 table of critical values is located in Appendix D. The χ^2 distribution is bounded on the left at zero (meaning that there are no negative values of χ^2) because it is a squared measure and therefore can take on only positive values. Like the t curve, the χ^2 distribution is a family of differently shaped curves, and each curve's shape is determined by degrees of freedom (df). As the df increases, the χ^2 distribution becomes more normal. Unlike the t curve, though, df for χ^2 are based not on sample size but, rather, on the size of the crosstabs table. Looking at Table 10.3, you can see that there are two rows (*female* and *male*) and two columns (*favor* and *oppose*). The **marginals** (row and column totals) are not included in the df calculation. The formula for degrees of freedom in a χ^2 distribution is

$$df = (r-1)(c-1), \text{where}$$

Formula 10(1)

r = the number of rows, excluding the marginal, and

c = the number of columns, excluding the marginal.

χ^2 distribution: The sampling/probability distribution for chi-square tests.

Marginals: Row and column totals in a bivariate contingency table.

Inserting the values from the gender and death penalty attitudes example into the formula, the result is

$$df = (2-1)(2-1) = (1)(1) = 1.$$

Step 3. Identify the Critical Value of the Test Statistic and State the Decision Rule

Remember in Chapter 8 when we used the α (alpha) level to find a particular value of z or t to plug into a confidence interval formula? You learned that the critical value is the number that cuts α off the tail of the distribution. If $\alpha = .05$, for instance, then the values of the test statistic that are out in the tail beyond the critical value constitute just 5% of the entire distribution. In other words, these values have a .05 or less probability of occurring. These values, then, are extremely unlikely to occur.

The process of finding the critical value of χ^2 (symbolized χ^2_{crit}) employs the same logic as that for finding critical values of z or t. The value of χ^2_{crit} depends on (1) the α level and (2) the df. Alpha must be set a priori so that the critical value can be determined before the test is run. Alpha can techni-cally be set at any number, but .05 and .01 are the most commonly used α levels in criminal justice/criminol-ogy. For the present example, we will choose $\alpha = .05$. Using Appendix D and finding the number at the intersection of $\alpha = .05$ and $df = 1$, it can be seen that $\chi^2_{crit} = 3.841$. This is the value that cuts .05 of the cases off the tail of the χ^2 distribution. Figure 10.1 illustrates this concept.

Figure 10.1 The Chi-Square Distribution, α, χ^2_{crit}, and χ^2_{obt}

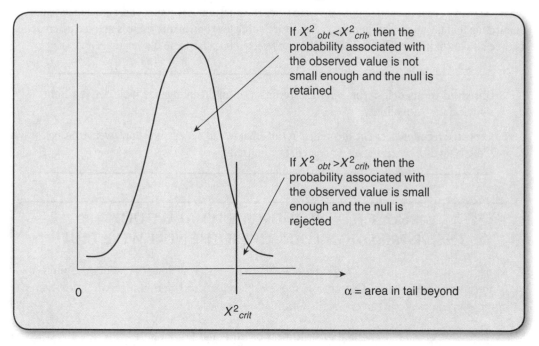

The decision rule is the a priori statement regarding the action you will take with respect to the null hypothesis based on the results of the statistical analysis that you are going to do in Step 4. The final product of Step 4 will be the **obtained value** of the test statistic, which in this case is the obtained value of χ^2 or, in shorthand, χ^2_{obt}. The null hypothesis will be rejected if the obtained value exceeds the critical value. If $\chi^2_{obt} > \chi^2_{crit}$, then the probability of obtaining this particular χ^2_{obt} value by chance alone is less than .05. Another way to think about it is that the probability of H_0 being true is less than .05. This is unlikely indeed! This would lead us to reject the null in favor of the alternative. The decision rule for the current test is: *If $\chi^2_{obt} > 3.841$, H_0 will be rejected.* This concept is depicted in Figure 10.1.

Obtained value: The value of the test statistic arrived at using the mathematical formulas specific to a particular test. The obtained value is the final product of Step 4 of a hypothesis test.

Step 4. Compute the Obtained Value of the Test Statistic

Now that we know the critical value, it is time to complete the analytical portion of the hypothesis test. Step 4 will culminate in the production of the obtained value, or χ^2_{obt}. In substantive terms, χ^2_{obt} is a measure of the difference between **observed frequencies** (f_o) and **expected frequencies** (f_e). Observed frequencies are the empirical values that appear in the crosstabs table produced from the sample-derived data set. Expected frequencies are the frequencies that *would* appear if the two variables under examination were unrelated to one another. In other words, the expected frequencies are what you would see if the null hypothesis were true. The question is whether observed equals expected (indicating that the null is true and the variables are unrelated) or whether there is marked discrepancy between them (indicating that the null should be rejected because there is a relationship).

Observed frequencies: The empirical results seen in a contingency table derived from sample data. Symbolized f_o.

Expected frequencies: The theoretical results that would be seen if the null were true, i.e., if the IV and DV were, in fact, unrelated. Symbolized f_e.

EXPECTED FREQUENCIES: WHAT WOULD THE DISTRIBUTION LOOK LIKE IF THE NULL WERE TRUE?

Consider the following two hypothetical distributions. The first represents the pattern of frequencies that would emerge if the independent and dependent variables were completely unrelated. The crosstabs table would look like this:

Hypothetical Distribution of Expected Frequencies

Independent Variable	Dependent Variable		
	X	Y	Row Marginal
A	25	25	50
B	25	25	50
Column Marginal	50	50	N = 100

You can see that the 100 cases in the sample are evenly distributed across the cells of the table. In a distribution like this, information about the independent variable (i.e., whether a particular case falls into class *A* or class *B*) offers no information whatsoever about the dependent variable (i.e., whether that case is in class *X* or class *Y*).

Now suppose you gathered data and constructed the following empirical/observed distribution:

Hypothetical Distribution of Observed Frequencies

Independent Variable	Dependent Variable		
	X	*Y*	*Row Marginal*
A	10	40	50
B	40	10	50
Column Marginal	50	50	*N* = 100

See the difference? Now the 100 cases are spread in a decidedly *uneven* manner across the cells of the table. It is clear in the observed distribution that cases in class *A* on the IV are very likely to be in class *Y* on the DV, while cases in the IV's class *B* are very likely to be in class *Y*. Knowing a case's placement on the independent variable thus *does* help predict its location on the dependent variable. This means that there is a statistical association between the IV and DV.

There are a few different steps in the process of computing χ^2_{obt}. Since this statistic is a measure of the difference between observed and expected frequencies, the first thing that must be done is to compute the latter. The formula for an expected frequency count is

$$f_{e_i} = \frac{rm_i \bullet cm_i}{N}, \text{where}$$

Formula 10(2)

f_{e_i} = the expected frequency for cell *i*,

rm_i = the row marginal of cell *i*,

cm_i = the column marginal of cell *i*,

N = the total sample size.

Every cell of the crosstabs table has an expected frequency, so the f_{e_i} formula must be completed for each one. To keep track of the cells, it is a good idea to label them. Notice in Table 10.3 that within each cell is a superscript with the letters A through D—these are the labels. Using Formula 10(2) for each cell:

$$f_{e_A} = \frac{1583 \bullet 1885}{2815} = 1060.02 \quad f_{e_B} = \frac{1583 \bullet 930}{2815} = 522.98$$

$$f_{e_C} = \frac{1232 \bullet 1885}{2815} = 824.98 \quad f_{e_D} = \frac{1232 \bullet 930}{2815} = 407.02.$$

Once the expected frequencies have been calculated, χ^2_{obt} can be computed using the formula:

$$\chi^2_{obt} = \Sigma \frac{\left(f_{o_i} - f_{e_i}\right)^2}{f_{e_i}}, \quad \text{where}$$

Formula 10(3)

f_{o_i} = the observed frequency of cell i,

f_{e_i} = the expected frequency of cell i.

Setting up a table can help you work through the process methodically. Table 10.4 contains the different elements of the χ^2_{obt} formula broken down across the columns.

The obtained value of the test statistic is found by summing the final column of the table, as such:

$$\chi^2_{obt} = 2.36 + 4.78 + 3.03 + 6.15 = 16.32.$$

Table 10.4   Calculating χ^2_{obt}

Cell	f_{oi}	f_{ei}	$(f_{oi} - f_{ei})$	$(f_{oi} - f_{ei})^2$	$\dfrac{\left(f_{o_i} - f_{e_i}\right)^2}{f_{e_i}}$
A	1010	1060.02	−50.02	2502.00	2.36
B	573	522.98	50.02	2502.00	4.78
C	875	824.98	50.02	2502.00	3.03
D	357	407.02	−50.02	2502.00	6.15
	$N = 2815$	$N = 2815$	$\Sigma = .00$		

Step 5. Make a Decision About the Null and State the Substantive Conclusion

It is time to decide whether to retain or reject the null. To do this, revisit the decision rule laid out in Step 3. It was stated that the null would be rejected if the obtained value of the test statistic exceeded 3.841. The obtained value turned out to be 16.32, which means that $\chi^2_{obt} > \chi^2_{crit}$, so the null is rejected. The alternative hypothesis is what we will take as being the true state of affairs.

The final stage of hypothesis testing is to interpret the results. People who conduct statistical analyses are responsible for communicating their findings in a manner that effectively resonates with their audience, be that audience composed of scholars, practitioners, the general public, or the media. It is especially important when discussing statistical findings with lay audiences that clear explanations be provided about what a set of quantitative results actually *means* in a substantive, practical sense. This makes findings accessible to a wide array of audiences who may find criminological results interesting and useful.

In the context of the present example, rejecting the null leads to the conclusion that the IV and DV are statistically related. More formally, we can state that there is a statistically significant relationship between people's gender and their attitudes about the death penalty. Note the language used in this conclusion—there is no cause-and-effect assertion being advanced. This is because the relationship that seems to be present in this bivariate analysis may actually be the result of unmeasured omitted variables that are the real driving force behind the gender differences. We have not, for instance, measured age, race, political beliefs, or religiosity, all of which may relate to people's beliefs about the effectiveness and morality of capital punishment. If women differ from men systematically on any of these characteristics, then the gender-attitude relationship might be spurious, meaning it is an artificial association that is actually the product of a third (or fourth) variable that has not been accounted for in the analysis. It is best to keep your language toned down and to use words like *relationship* and *association* rather than *cause* or *effect*.

Another Example

For the second example, we will use the juvenile race/ethnicity and sentencing data displayed in Table 10.2 and reproduced in the table below. The data came from the BJS Juvenile Defendants in State Criminal Courts (JDCC; see Data Sources 10). Using an alpha level of .01, we will test for a relationship between the IV (race/ethnicity) and the DV (sentence type). All five steps will be used.

Contingency Table for Juvenile Race or Ethnicity and Sentence Received

| Race or Ethnicity | *Sentence Received* | | | |
	Jail	Prison	Probation	Row Marginal
White	226[A]	102[B]	333[C]	661
Black	398[D]	295[E]	917[F]	1,610
Hispanic	177[G]	83[H]	195[I]	455
Column Marginal	801	480	1,445	$N = 2,726$

DATA SOURCES 10.1

JUVENILE DEFENDANTS IN CRIMINAL COURTS (JDCC)

The JDCC is a subset of the Bureau of Justice Statistics' (BJS) State Court Processing series that gathers information on defendants convicted of felonies in large, urban counties. BJS researchers pulled information about juveniles charged with felonies in 40 of these counties in May 1998. Each case was tracked through disposition. Information about the juveniles' demographics, court processes, final dispositions, and sentences was recorded. Due to issues with access to and acquisition of data in some of the counties, the JDCC is a nonprobability sample, and conclusions drawn from it should therefore be interpreted cautiously (Bureau of Justice Statistics, 1998).

Step 1. State the Null (H_0) and Alternative (H_1) Hypotheses

$$H_0: \chi^2 = 0$$

$$H_1: \chi^2 > 0$$

Step 2. Identify the Distribution and Compute the Degrees of Freedom

The distribution is χ^2 and the $df = (3 - 1)(3 - 1) = (2)(2) = 4$.

Step 3. Identify the Critical Value of the Test Statistic and State the Decision Rule

With $\alpha = .01$ and $df = 4$, $\chi^2_{crit} = 13.277$. The decision rule is: *If $\chi^2_{obt} > 13.277$, H_0 will be rejected.*

Step 4. Compute the Obtained Value of the Test Statistic

First, we need to calculate the expected frequencies using Formula 10(3). The frequencies for the first three cells (labeled A, B, and C, left to right) are:

$$f_{e_A} = \frac{661 \bullet 801}{2726} = 194.23 \quad f_{e_B} = \frac{661 \bullet 480}{2726} = 116.39 \quad f_{e_C} = \frac{661 \bullet 1445}{2726} = 350.38$$

Can you compute the rest? Try it out as practice.

Next, the computational table is used to calculate χ^2_{obt} (Table 10.5).

Table 10.5 Calculating χ^2_{obt}

Cell	f_{o_i}	f_{e_i}	$(f_{o_i} - f_{e_i})$	$(f_{o_i} - f_{e_i})^2$	$\dfrac{\left(f_{o_i} - f_{e_i}\right)^2}{f_{e_i}}$
A	226	194.23	31.77	1009.33	5.20
B	102	116.39	−14.39	207.07	1.78
C	333	350.38	−17.38	302.06	.86
D	398	473.08	−75.08	5637.01	11.92
E	295	283.49	11.51	132.48	.47
F	917	853.43	63.57	4041.14	4.74
G	177	133.70	43.30	1874.89	14.02
H	83	80.12	2.88	8.29	.10
I	195	241.19	−46.19	2133.52	8.85
	$N = 2726$	$N = 2726.01$	$\Sigma = -.01$		$\Sigma = 47.94$

The obtained value of the test statistic is 47.94.

Step 5. Make a Decision About the Null and State the Substantive Conclusion

Since χ^2_{obt} (47.94) is greater than 13.277, the null is rejected. There is a statistically significant relationship between juveniles' race/ethnicity and the sentences they receive. Remember that no claim is being made that race is the driving force behind sentencing decisions or that any particular racial group is uniformly treated better or worse than others. Conclusions must be phrased in a more careful, circumspect manner because there are only two variables being analyzed. In the present example, we have not accounted for the important influence that offense severity and offenders' prior criminal records exert on the sentencing decision. If there are racial differences in offense severity or criminal history, then the race-sentencing relationship could be spurious.

▣ MEASURES OF ASSOCIATION

The chi-square test alerts you when there is a statistically signficant relationship between two variables, but it is silent as to the strength or magnitude of that relationship. We know from the above two examples, for instance, that gender is related to attitudes toward capital punishment and that convicted juveniles' race/ethnicity is related to the sentences they receive, but we do not know the magnitudes of these associations: They could be very strong, moderate, or quite weak. This question is an important one because a very slight relationship—even if statistically significant in a technical sense—is not of much substantive or practical importance. Large relationships are more meaningful in the "real world."

There are several **measures of association,** and this chapter covers four of them. The level of measurement of the IV and DV dictate which measures are appropriate for a given analysis. It is important to keep in mind that measures of association are computed only when the null hypothesis has been rejected—if the null is not rejected and you conclude that there is no relationship between the IV and DV, then it makes no sense to go on and try to interpret an association you just said does not exist. The following discussion will introduce four tests (two for nominal and two for ordinal data), and then the next section will show you how to use SPSS to compute χ^2_{obt} and accompanying measures of association.

Measures of association: Procedures for determining the strength or magnitude of a relationship between an IV and a DV. Used only when the null hypothesis has been rejected.

Nominal Data: Lambda and Cramer's *V*

When the variables under examination are nominal, two measures of association that can be used to assess the magnitude of a statistically significant relationship are **lambda** and **Cramer's *V***. Both of these measures range from 0.00 to 1.00, with higher values indicative of stronger IV-DV associations. Both of them also have known sampling distributions and their statistical significance can thus be computed. We will see an illustration of this in the SPSS section.

Lambda: An asymmetrical measure of association for χ^2 when the variables are nominal. Lambda ranges from 0.00 to 1.00 and is a proportionate reduction in error measure.

Cramer's *V*: A symmetrical measure of association for χ^2 when the variables are nominal. *V* ranges from 0.00 to 1.00 and indicates the strength of the IV-DV relationship. Higher values represent stronger relationships.

Lambda is a *proportionate reduction in error* (PRE) measure. Proportionate reduction in error refers to the extent to which knowing a person or object's placement on an IV helps predict that person or object's classification on the DV. Consider the gender and death penalty attitudes example from above. If you were trying to predict a given individual's attitudes toward capital punishment and the only piece of information you had was the frequency distribution of this DV (that is, you knew that 1,885 people in the sample support capital punishment and 930 oppose it), then your best bet would be to guess the modal category (support) because that is the guessing strategy that would produce the fewest prediction errors. There would, though, be a substantial number of these errors—930, to be exact! Now, suppose that you know a given person's gender. To what extent does this knowledge improve your accuracy when you predict whether that person opposes or favors the death penalty? This is the idea behind PRE measures like lambda.

Lambda is a proportion. A lambda of .60, for instance, would mean that knowing people's classifications on the IV reduces DV prediction errors by 60%. This is a respectable reduction in error and signals that the IV is a strong predictor of the DV. A lambda of .05, by contrast, would represent an error reduction of just 5%, which is very poor and indicates a weak relationship.

Lambda is an asymmetrical measure, meaning that you have to be mindful about which variable has been designated the IV and which the DV. The value of lambda changes if the variables' order is reversed. If there is no clear IV and DV designation, then you should forego lambda and use Cramer's *V* instead.

Cramer's *V* is a symmetrical measure that ranges from 0.00 to 1.00. Since it is symmetrical, *V* always takes on the same value regardless of which variable is called the IV and which the DV. Higher values of *V* indicate stronger relationships. A *V* of .10, for instance, would signify a weak relationship, while a value of .80 would represent a strong relationship.

Ordinal Data: Gamma and Somers' *d*

When data are ordinal, **Goodman and Kruskal's gamma** and **Somers' *d*** are good statistics for measuring the strength of associations. These statistics both range from −1.00 to 1.00, with zero meaning no relationship, −1.00 indicating a perfect negative relationship (as the IV increases, the DV decreases), and 1.00 representing a perfect positive relationship (as the IV increases, so too does the DV). Gamma is symmetrical and *d* is asymmetrical. Both have sampling distributions and it can therefore be determined whether or not they are significantly different from zero.

Goodman and Kruskal's gamma: A symmetrical measure of association for χ^2 when the variables are ordinal. Gamma ranges from −1.00 to +1.00.

Somers' *d*: An asymmetrical measure of association for χ^2 when the variables are nominal. Somers' *d* ranges from −1.00 to +1.00.

🔲 SPSS

The SPSS program can be used to generate χ^2_{obt} and accompanying measures of association. The chi-square analysis is found via the sequence *Analyze* → *Descriptive Statistics* → *Crosstabs*. Let us first consider the gender and capital punishment example from above. Figure 10.2 shows the dialog boxes involved in running this analysis in SPSS. Note that you must check the box labeled *Chi-square* in order to get a chi-square analysis; if you do not check this box, SPSS will merely give you a crosstabs table. Lambda and Cramer's *V* are the appropriate measures of association because both of these variables are nominal. Figure 10.3 shows the first portion of the output, which displays the results of the χ^2 test.

The obtained value of the χ^2 statistic is located on the line labeled *Pearson Chi-Square*. You can see in Figure 10.3 that $\chi^2_{obt} = 16.324$, which is identical to the value we obtained by hand. The output also tells you whether or not the null should be rejected, but it does so in a way that we have not seen before. SPSS gives you what is called a ***p* value**. *P* values tell you the exact probability of the obtained value of the test statistic. The *p* value in SPSS χ^2 output is the number located at the intersection of the *Asymp. Sig. (2-sided)* column and the *Pearson Chi-Square* row. Here, $p = .000$. What you do is compare *p* to α.

If p is less than α, the null is rejected; if p is greater than α, the null is retained. Since in this problem α was set at .05, the null hypothesis is rejected because .000 < .05. There is a statistically significant relationship between gender and death penalty attitudes.

Figure 10.2   Running a Chi-Square Test and Measures of Association in SPSS

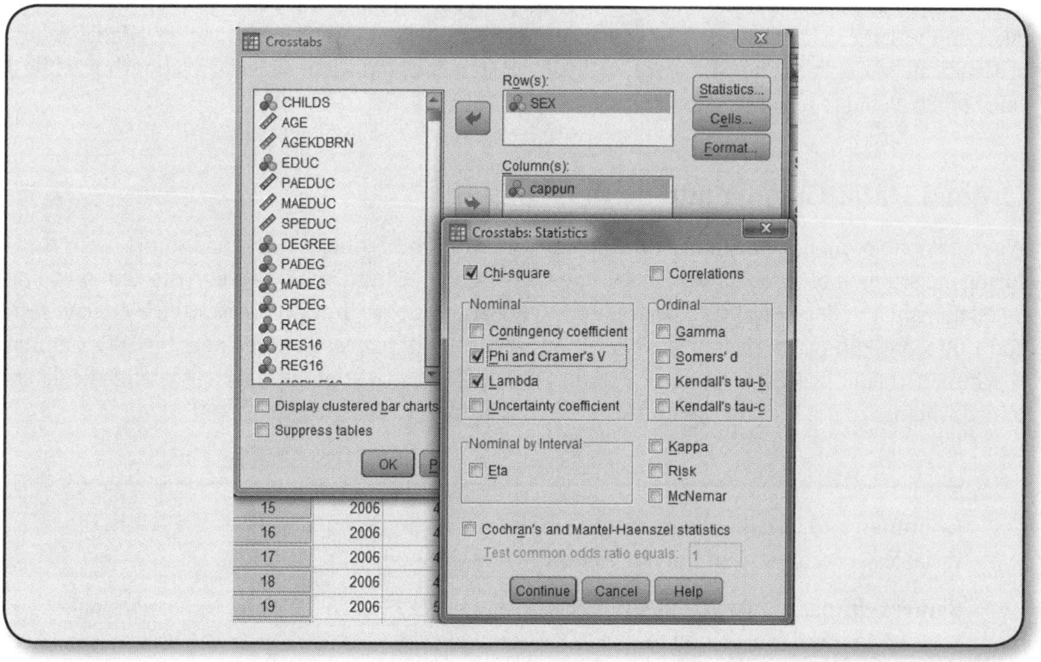

Figure 10.3   Chi-Square Output

Chi-Square Tests

	Value	df	Asymp. Sig. (2-sided)	Exact Sig. (2-sided)	Exact Sig. (1-sided)
Pearson Chi-Square	16.324[a]	1	.000		
Continuity Correction[b]	15.999	1	.000		
Likelihood Ratio	16.426	1	.000		
Fisher's Exact Test				.000	.000
Linear-by-Linear Association	16.318	1	.000		
N of Valid Cases	2815				

a. 0 cells (.0%) have expected count less than 5. The minimum expected count is 407.02.

b. Computed only for a 2x2 table

***p* value:** In SPSS output, the probability associated with the obtained value of the test statistic. When $p < \alpha$, the null hypothesis is rejected.

As you know, though, rejection of the null hypothesis is only part of the story because the χ^2 statistic does not offer information about the magnitude or strength of the IV-DV relationship. For this, we turn to measures of association. See Figure 10.4, where the values of lambda and Cramer's *V* are reported.

Figure 10.4   Measures of Association

Directional Measures

			Value	Asymp. Std. Error[a]	Approx. T	Approx. Sig.
Nominal by Nominal	Lambda	Symmetric	.000	.000	[b]	[b]
		RESPONDENTS SEX Dependent	.000	.000	[b]	[b]
		FAVOR OR OPPOSE DEATH PENALTY FOR MURDER Dependent	.000	.000	[b]	[b]
	Goodman and Kruskal tau	RESPONDENTS SEX Dependent	.006	.003		.000[c]
		FAVOR OR OPPOSE DEATH PENALTY FOR MURDER Dependent	.006	.003		.000[c]

a. Not assuming the null hypothesis.

b. Cannot be computed because the asymptotic standard error equals zero.

c. Based on chi-square approximation

Symmetric Measures

		Value	Approx. Sig.
Nominal by Nominal	Phi	.076	.000
	Cramer's V	.076	.000
N of Valid Cases		2815	

Judging both by lambda and by Cramer's *V*, this relationship is very weak. Lambda is zero, which means that knowing GSS respondents' gender does not reduce the number of errors made in predicting their death penalty attitudes. *V* is .076, which is statistically significant (as evidenced by the fact that the *Approx. Sig.* value is .000, which is less than .05) but nonetheless practically zero. This demonstrates how statistical significance can be misleading—a relationship might be significant in a technical sense but still very *in*significant in practical terms.

Let us try a chi-square analysis with the ordinal variables from Table 10.1a. The question in that example was whether people's experiences as the subjects of recent traffic stops would affect their willingness to call the police for assistance. We will use an alpha level of .01. Following the same procedures pictured in Figure 10.2 but choosing Somers' d and gamma as the measures of association, the output in Figure 10.5 is obtained.

Figure 10.5   SPSS Chi-Square Output for Traffic Stops and Reporting Neighborhood Problems

Chi-Square Tests

	Value	df	Asymp. Sig. (2-sided)
Pearson Chi-Square	340.328[a]	2	.000
Likelihood Ratio	352.675	2	.000
Linear-by-Linear Association	260.908	1	.000
N of Valid Cases	14966		

a. 0 cells (.0%) have expected count less than 5. The minimum expected count is 143.98.

Directional Measures

			Value	Asymp. Std. Error[a]	Approx. T[b]	Approx. Sig.
Ordinal by Ordinal	Somers' d	Symmetric	-.117	.006	-17.685	.000
		Number of times respondent has been in vehicle stopped by police Dependent	-.245	.013	-17.685	.000
		Number of times respondent has called police to report neighborhood problem Dependent	-.077	.004	-17.685	.000

a. Not assuming the null hypothesis.
b. Using the asymptotic standard error assuming the null hypothesis.

Symmetric Measures

		Value	Asymp. Std. Error[a]	Approx. T[b]	Approx. Sig.
Ordinal by Ordinal	Gamma	-.439	.024	-17.685	.000
N of Valid Cases		14966			

a. Not assuming the null hypothesis.
b. Using the asymptotic standard error assuming the null hypothesis.

The obtained value (340.328) is statistically significant (.000 < .01), so there does appear to be a relationship between the two variables. Somers' d, though, suggests that the relationship is weak in magnitude; at only −.077, the relationship is very near zero. Gamma implies that the association is somewhat moderate, as $-.439$ is nearly halfway between zero and -1.0. The negative signs indicate that this is an inverse relationship; that is, higher values on the IV are associated with lower values on the DV. People who have experienced one or more recent traffic stops are less likely to have called the police in the past year to report neighborhood problems. There does, then, appear to be some support for the hypothesis that being stopped by police affects people's willingness to call the police for help or assistance, but this support is tempered by the fact that the relationship is not very strong.

CHAPTER SUMMARY

This chapter introduced the chi-square test of independence, which is the hypothesis-testing procedure appropriate when both of the variables under examination are categorical. The key elements of the χ^2 test are observed and expected frequencies. Observed frequencies are the empirical results seen in the sample, while expected frequencies are those that would appear if the null hypothesis were true and the two variables unrelated. χ^2_{obt} is a measure of the difference between observed and expected, and comparing χ^2_{obt} to χ^2_{crit} for a set α level allows for a determination of whether the null hypothesis should be retained or rejected.

When the null is retained (that is, when $\chi^2_{obt} < \chi^2_{crit}$), the substantive conclusion is that the two variables are not related. When the null is rejected (when $\chi^2_{obt} > \chi^2_{crit}$), the conclusion is that there is a relationship between them. *Statistical* significance, though, is only a necessary and not a sufficient condition for *practical* significance. The chi-square statistic does not offer information about the strength of a relationship and how substantively meaningful this association actually is.

For this, measures of association are turned to when the null has been rejected. For nominal variables, lambda and Cramer's V are useful, while for ordinal variables, Somers' d and gamma can be employed. SPSS can be programmed to provide chi-square tests and measures of association. You should always generate measures of association when you run χ^2 tests yourself, and you should always expect them from other people who run these analyses and present you with the results. Statistical significance is important, but the magnitude of the relationship tells you just how meaningful the association is in practical terms.

CHAPTER 10 REVIEW PROBLEMS

1. A primary argument advanced in favor of the death penalty is that it is a necessary tool for controlling violent crime. Based on this rationale and the fact that some states currently authorize capital punishment by law and some do not, it could be hypothesized that states with higher violent crime rates are more likely to authorize the death penalty as compared to states with lower violent crime rates. The following table contains violent crime data from the 2009 Uniform Crime Reports and measures whether states were above or below the median violent crime rate during this year. It also shows data from the Bureau of Justice Statistics regarding whether or not states authorize execution as a criminal punishment for murder.
   a. Identify the independent and dependent variables.
   b. Identify the level of measurement of each variable.
   c. Using an alpha level of .01, conduct a five-step chi-square hypothesis test to determine whether or not the two variables are related. Use all five steps.

Violent Crime Rate	State Has Death Penalty?		Row marginal
	Yes	No	
High	22^A	4^B	26
Low	14^C	10^D	24
Column marginal	36	14	$N = 50$

2. Criminal defendants who wish to exercise their right to trial rather than dispose of their case by pleading guilty are often permitted to choose between a jury trial and a bench trial (a trial before a judge rather than a jury). A question stemming from this is whether there is a relationship between trial type and sentencing; specifically, are defendants more or less likely to be sentenced to prison depending on whether they were convicted by a jury or a judge? To speak to this question, the table below contains data from the 2004 National Judicial Reporting Program.

Trial Type	Sentenced to Prison?		Row marginal
	Yes	No	
Jury	167^A	251^B	418
Bench	88^C	320^D	408
Column marginal	255	571	$N = 826$

a. Identify the independent and dependent variables.
b. Identify the level of measurement of each variable.
c. Using an alpha level of .05, conduct a five-step chi-square hypothesis test to determine whether or not the two variables are related. Use all five steps.

3. One criticism of racial profiling studies is that people's driving frequency is often unaccounted for. This is a problem because all else being equal, people on the road often are more likely to get pulled over simply because they have greater opportunities for committing traffic infractions and they spend more time on public roads where police may be watching. The table below contains 2004 Police-Public Contact Survey (PPCS) data narrowed down to black male respondents. The variables measure driving frequency and whether or not these respondents had been stopped by police for traffic offenses within the past 12 months.

Driving Frequency	Experienced a Traffic Stop?		Row marginal
	Yes	No	
Almost Every Day	200^A	139^B	339
Often	13^C	17^D	30
Rarely	12^E	13^F	25
Column marginal	225	169	$N = 394$

a. Identify the independent and dependent variables.
b. Identify the level of measurement of each variable.
c. Using an alpha level of .01, conduct a five-step chi-square hypothesis test to determine whether or not the two variables are related. Use all five steps.

4. In the second example presented in the preceding chapter, we saw that transferred juveniles' race/ethnicity was significantly related to the sentences they received. It was also noted in the text, however, that the race-sentencing relationship might be affected by juveniles' prior criminal history and, in particular, whether juveniles of any particular race or races are more likely to have criminal records relative to juveniles of other races. The table below contains Juvenile Defendants in Criminal Court data with the same race/ethnicity variable and a variable measuring whether or not the youth defendants possessed histories of juvenile arrests or convictions.

| *Race* | Prior Juvenile Arrests or Convictions? | | |
	Yes	*No*	*Row marginal*
Black	1625^A	506^B	2131
White	470^C	191^D	661
Hispanic	459^E	147^F	606
Column marginal	2554	844	$N = 3{,}398$

a. Identify the independent and dependent variables.
b. Identify the level of measurement of each variable.
c. Using an alpha level of .05, conduct a five-step chi-square hypothesis test to determine whether or not the two variables are related. Use all five steps.

5. The data from Table 10.1b are reproduced below.

| *Number of Stops* | Number of Times Called to Report Problems | | |
	None	*One or More*	*Row Marginal*
None	$6{,}409^A$	$1{,}095^B$	7,504
One	$5{,}724^C$	323^D	6,047
More than One	$1{,}335^E$	92^F	1,427
Column Marginal	13,468	1,510	$N = 14{,}978$

a. Identify the independent and dependent variables.
b. Identify the level of measurement of each variable.
c. Using an alpha level of .001, conduct a five-step chi-square hypothesis test to determine whether or not the two variables are related. Use all five steps.

6. Is there a relationship between the gender of a victim of firearm violence and the type of firearm used in the assault? The Firearm Injury Surveillance Study provides information on injury intent, victim gender, and gun type. For this analysis, the sample has been narrowed to emergency department patients during the years 2005 to 2007 whose injuries were the result of assaults.

Victim Gender	Firearm Type			Row marginal
	Handgun	Rifle	Shotgun	
Male	2,347[A]	54[B]	174[C]	2,575
Female	274[D]	8[E]	25[F]	307
Column marginal	2,621	62	199	$N = 2,882$

a. Identify the independent and dependent variables.
b. Identify the level of measurement of each variable.
c. Using an alpha level of .05, conduct a five-step chi-square hypothesis test to determine whether or not the two variables are related. Use all five steps.

7. One criticism of private prisons is that they are profit-driven and therefore have an incentive to cut corners wherever possible. This may cause threats to institutional security and reduced opportunities for rehabilitation and self-improvement among persons incarcerated in these facilities. The Census of State and Federal Correctional Facilities data set contains information about whether a prison is publicly or privately operated and whether that institution offers college courses to its inmates. The table below contains data from a random sample of prisons.

Facility Type	Prison Offers College Courses?		Row marginal
	Yes	No	
Private	79[A]	336[B]	415
Public	562[C]	840[D]	1,402
Column marginal	641	1,176	$N = 1,817$

a. Identify the independent and dependent variables.
b. Identify the level of measurement of each variable.
c. Using an alpha level of .01, conduct a five-step chi-square hypothesis test to determine whether or not the two variables are related. Use all five steps.

8. The table below contains data from the same sample of prisons used in Question 7, with the question being whether private prisons differ significantly from public ones in terms of the offering of vocational programs to inmates.

Facility Type	Prison Offers Vocational Training?		Row marginal
	Yes	No	
Private	109[A]	306[B]	415
Public	846[C]	556[D]	1,402
Column marginal	955	862	$N = 1,817$

a. Identify the independent and dependent variables.
b. Identify the level of measurement of each variable.

c. Using an alpha level of .001, conduct a five-step chi-square hypothesis test to determine whether or not the two variables are related. Use all five steps.

9. The relationship between gender and crime has been studied extensively. It is well-documented that females have a lower crime commission rate than males do and, moreover, that females' crimes tend to be nonviolent, such as shoplifting and prostitution. Let us test for a relationship between gender and crime using the State Court Processing Statistics data set. The crosstabs table below shows the gender of convicted defendants and the felony types for which they were ultimately convicted.

Defendant Gender	Felony Type				
	Violent	*Property*	*Drug*	*Other*	*Row marginal*
Male	1,302^A	1961^B	2,540^C	974^D	6,777
Female	155^E	557^F	498^G	118^H	1,328
Column marginal	1,457	2,518	3,038	1,092	$N = 8,105$

a. Identify the independent and dependent variables.
b. Identify the level of measurement of each variable.
c. Using an alpha level of .05, conduct a five-step chi-square hypothesis test to determine whether or not the two variables are related. Use all five steps.

10. When researchers are analyzing the relationship between defendants' race and the sentences they receive, it is important to control for factors such as prior criminal history. Racial differences in criminal history could impact the race-sentencing relationship. The table below contains race and prior felony conviction data from the State Court Processing Statistics.

Defendant Race/Ethnicity	Number of Prior Felony Convictions			
	0	*1 – 9*	*10+*	*Row marginal*
White	1,025^A	642^B	43^C	1,710
Black	1,011^D	778^E	60^F	1,849
Hispanic	663^G	354^H	14^I	1,031
Column marginal	2,699	1,774	117	$N = 4,590$

a. Identify the independent and dependent variables.
b. Identify the level of measurement of each variable.
c. Using an alpha level of .05, conduct a five-step chi-square hypothesis test to determine whether or not the two variables are related. Use all five steps.

11. Some studies have shown that men are more likely than women to express negative attitudes toward police. The table below contains PPCS data for white respondents who reported having been the subject of a traffic stop in the past year. The variable *legitimate* measures whether or not respondents believed that police had a legitimate reason for pulling them over.

a. Identify the independent and dependent variables.
b. Identify the level of measurement of each variable.

c. Using an alpha level of .05, conduct a five-step chi-square hypothesis test to determine whether or not the two variables are related. Use all five steps.

Respondent Gender	Legitimate Reason for Stop?		
	No	Yes	Row marginal
Male	263[A]	1,667[B]	1,930
Female	166[C]	1,237[D]	1,403
Column marginal	429	2,904	$N = 3,333$

12. Does the gender-legitimacy relationship vary by race? The table below contains PPCS data for black respondents who reported having been the subject of a traffic stop in the past year. The variable *legitimate* measures whether or not respondents believed that police had a legitimate reason for pulling them over.

Respondent Gender	Legitimate Reason for Stop?		
	No	Yes	Row marginal
Male	45[A]	139[B]	184
Female	38[C]	134[D]	172
Column marginal	83	273	$N = 356$

a. Identify the independent and dependent variables.
b. Identify the level of measurement of each variable.
c. Using an alpha level of .05, conduct a five-step chi-square hypothesis test to determine whether or not the two variables are related. Use all five steps.

13. Is there a relationship between defendants' gender and their likelihood of pleading guilty or choosing to go to trial? The companion website (**http://www.sagepub.com/gau**) contains the SPSS data file *State Court Processing Statistics for Chapter 10.sav.* Use SPSS to run a chi-square analysis to test for independence between *gender* and plea. Based on the variables' level of measurement, select appropriate measures of association. Then do the following.

a. Identify the obtained value of the chi-square statistic.
b. Make a decision about whether or not you would reject the null hypothesis of independence at an alpha level of .05 *and* explain how you arrived at that decision.
c. State the conclusion that you draw from the results of each of these analyses in terms of whether or not there is a relationship between people's experiences with traffic stops and their likelihood of calling the police for assistance.
d. *If you rejected the null hypothesis,* interpret the measures of association. How strong is the relationship between the IV and each DV? Would you say that this is a substantively meaningful relationship?

14. Part of the debate about racial profiling in traffic stops revolves around police actions during stops; specifically, some people argue that police are more likely to engage in pretextual stops with minority drivers. Pretext stops occur when an officer uses a traffic infraction as a means of getting close to a vehicle or driver, often with the intent of asking the driver for permission to search the vehicle. The companion website (**http://www.sagepub.com/gau**) contains variables from the 2005 Police–Public Contact Survey. The variables are *driver race* and *consent*, the latter of which measures whether or not police requested permission to search drivers' vehicles during traffic stops. Use SPSS to run a chi-square analysis. Based on the variables' level of measurement, select appropriate measures of association. Then do the following:

a. Identify the obtained value of the chi-square statistic.
b. Make a decision about whether or not you would reject the null hypothesis of independence at an alpha level of .01 *and* explain how you arrived at that decision.
c. State the conclusion that you draw from the results of each of these analyses as to whether or not there is a difference between private and public prisons in terms of offering vocational training.
d. *If you rejected the null hypothesis,* interpret the measures of association. How strong is the relationship between the IV and each DV? Would you say that this is a substantively meaningful relationship?

15. Is the security level of a prison related to the number of inmate escapes in that institution? The data file *Census of State and Federal Correctional Facilities for Chapter 10.sav* (**http://www.sagepub.com/gau**) contains the variables *security* and *escapes*. Use SPSS to run a chi-square analysis. Based on the variables' level of measurement, select appropriate measures of association. Then do the following:

a. Identify the obtained value of the chi-square statistic.
b. Make a decision about whether or not you would reject the null hypothesis of independence at an alpha level of .01 *and* explain how you arrived at that decision.
c. State the conclusion that you draw from the results of each of these analyses in terms of whether or not there is a difference between private and public prisons in terms of offering vocational training.
d. *If you rejected the null hypothesis,* interpret the measures of association. How strong is the relationship between the IV and each DV? Would you say that this is a substantively meaningful relationship?

KEY TERMS

Chi-square test of independence	Observed frequencies	Cramer's *V*
Statistical independence	Expected frequencies	Goodman and Kruskal's gamma
Statistical dependence	Measures of association	Somers' *d*
Obtained value	Lambda	*p* value

GLOSSARY OF SYMBOLS INTRODUCED IN THIS CHAPTER

χ^2	The Greek letter chi squared; a symbol for the test of independence and its associated sampling distribution
α	The Greek letter alpha, symbolizing the critical area in the tail of the distribution
χ^2_{crit}	The critical value of the chi-square statistic
f_o	Observed frequencies
f_e	Expected frequencies
rm	Row marginals; used for the calculation of expected frequencies
cm	Column marginals; used for the calculation of expected frequencies
χ^2_{obt}	The obtained value of the chi-square statistic
p	The obtained probability on SPSS output that is compared to the alpha level in order to make a decision about the null hypothesis

ENDNOTES

1. The raw numbers presented in this chapter were taken from the 1999 Police-Public Contact Survey. All analyses conducted using these data are original and do not reflect the research, opinion, or input of any of the authors of the Gibson et al. (2010) article.
2. The raw numbers presented in this chapter were taken from the 1998 Juvenile Defendants in Criminal Courts data set. All analyses conducted using these data are original and do not reflect the research, opinion, or input of the authors of the Jordan and Freiburger (2010) article.

REFERENCES

Bureau of Justice Statistics. (1998). *Juvenile Defendants in Criminal Court, 1998* [data file]. Washington, D.C.: U.S. Department of Justice.

Gibson, C. L., Walker, S., Jennings, W. G., & Miller, J. M. (2010). The impact of traffic stops on calling the police for help. *Criminal Justice Policy Review, 21*(2), 139–159.

Jordan, K. L., & Freiburger, T. L. (2010). Examining the impact of race and ethnicity on the sentencing of juveniles in the adult court. *Criminal Justice Policy Review, 21*(2), 185–201.

Chapter 11

Hypothesis Testing With Two Population Means or Proportions

There are many situations in which criminal justice/criminology researchers work with categorical independent variables (IVs) and continuous dependent variables (DVs). Research Example 11 illustrates one such instance.

RESEARCH EXAMPLE 11.1

Do Multiple Homicide Offenders Specialize in Killing?

Serial killers and mass murderers capture the public's curiosity and imagination. Who can resist some voyeuristic gawking at a killer who periodically snuffs out innocent victims while outwardly appearing to be a regular guy, or the tormented soul whose troubled life ultimately explodes in an episode of wanton slaughter? Popular portrayals of multiple homicide offenders (MHOs) lend the impression that these killers are fundamentally different from more ordinary criminals and from single homicide offenders (SHOs) in that they only commit homicide and lead otherwise crime-free lives. But is this popular conception true?

Wright, Pratt, and DeLisi (2008) decided to find out. They constructed an index measuring diversity of offending within a sample of homicide offenders. This index captured the extent to which homicide offenders committed only homicide and no other crimes versus the extent to which they engaged in various types of illegal acts. The researchers divided the sample into MHOs and SHOs and calculated each group's mean and standard deviation on the diversity index. They found the statistics located in the following table.

(Continued)

(Continued)

Diversity Index Means and Standard Deviations for Multiple Homicide Offenders and Single Homicide Offenders

	MHOs		SHOs	
	$\bar{x}$	s	$\bar{x}$	s
Diversity index	.36	.32	.37	.33

Source: Adapted From Table 1 in Wright, Pratt, and DeLisi (2008).

In Research Example 10.1, Wright et al. (2008) had two groups (MHOs and SHOs), each with their own mean and standard deviation, and the goal was to find out whether the groups' means differ significantly from one another. A significant difference would indicate that MHOs and SHOs do indeed differ in the variety of crimes they commit, while no disparity in the means would imply that these two types of homicide offenders are equally diverse in offending. What did the researchers do to find out whether MHOs and SHOs have significantly different diversity indices?

The answer is that they conducted a two-population test for differences between means, or what is more commonly referred to as a *t* **test** because it relies on the *t* distribution. These types of tests are the subject of this chapter. We will also cover two-population tests for differences between proportions, which are conceptually similar to *t* tests but employ the *z* distribution.

---

t **test:** The test used with a two-class, categorical independent variable and a continuous dependent variable.

Classes: The categories or groups on a categorical variable.

Both of these test types are appropriate only when the IV is categorical with exactly two **classes** or groups. Examples of two-class, categorical IVs include gender (*male; female*) and political orientation (*liberal; conservative*). The two test types differ, though, in terms of their dependent variables: *t* tests require continuous DVs, while tests for differences between proportions are for use with binary, categorical DVs. The diversity index that Wright et al. (2008) used is continuous, which is the reason they were able to compute a mean and standard deviation, and the reason that a *t* test is the appropriate analytic strategy. In the review problems at the end of the chapter, you will conduct a *t* test to find out whether MHOs and SHOs differ significantly in offending diversity.

▣ TWO-POPULATION TESTS FOR DIFFERENCES BETWEEN MEANS: *t* TESTS

There are many situations in which people working in criminal justice/criminology would want to test for differences between two means. Someone might be interested in finding out whether offenders who are sentenced to a term of incarceration receive significantly different mean sentence lengths depending on whether they are male or female. A municipal police department might implement an innovative new policing strategy and want to know whether the program significantly reduced mean crime rates in the city. These and a myriad of other scenarios in both academic and policy arenas require the use of *t* tests.

Sampling distributions are theoretical curves created when infinite samples are drawn from a single population and a statistic is computed and plotted for each sample. Over time, the distribution of sample statistics builds up and, if the sample size is large (i.e., $N \geq 100$), the statistics form a normal curve. There are also sampling distributions for differences between means. See Figure 11.1. These distributions center on the true population difference, $\mu_1 - \mu_2$.

A sampling distribution of differences between means is created by pulling infinite *pairs* of samples—rather than single samples—and in each pair, computing the two means, subtracting one mean from the other to form a difference score, and then plotting that difference score. Over time, the difference scores build up. If $N \geq 100$, the sampling distribution of differences between means is normal; if $N \leq 99$, then the distribution is more like "normalish" because it tends to become wide and flat. The *t* distribution, being flexible and able to accommodate various sample sizes, is the probability distribution of choice for tests of differences between means.

Figure 11.1 The Sampling Distribution of Differences Between Means

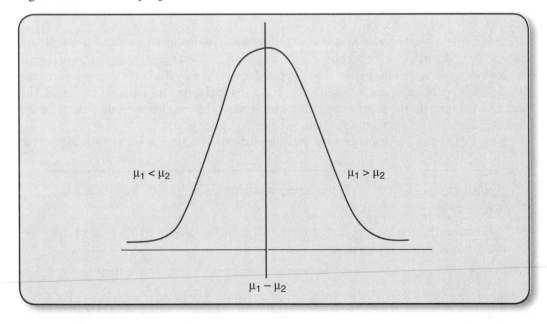

There are two general types of *t* tests, one for use with **independent samples** and one for use with **dependent samples**. The difference between them pertains to the method used to select the two samples under examination. In independent sampling designs, the selection of cases into one sample in no way affects or is affected by the selection of cases into the other sample. In dependent samples designs, by contrast, the two samples are somehow related to each other. The two major types of dependent samples designs are matched pairs and repeated measures.

Independent samples: Pairs of samples in which the selection of people or objects into one sample in no way affected or was affected by the selection of people or objects into the other sample.

Dependent samples: Pairs of samples in which the selection of people or objects into one sample directly affected or was directly affected by the selection of people or objects into the other sample. The most common types are matched pairs and repeated measures.

Matched-pairs designs are used when researchers need an experimental group and a control group but are unable to use random assignment to create the groups. They therefore gather a sample from a treatment group and then construct a control group via the deliberate selection of cases that did not receive the treatment but that are similar to the treatment group cases on key characteristics. If the unit of analysis is people, participants in the control group might be matched to the treatment group on race, gender, age, and criminal history.

Repeated measures designs are a common method used to evaluate program impact. These are before-and-after designs wherein the experimental group is measured prior to the intervention of interest and then again afterward to determine whether the postintervention scores differ significantly from the preintervention ones. In repeated measures, then, the two samples are actually the same people or objects measured twice.

The first step in deciding what kind of *t* test to use, then, is to figure out whether the samples are independent or dependent. If they are dependent, then the dependent samples *t* test is appropriate. If they are independent, then the independent samples *t* test is the proper analytic method. There are, though, two types of independent samples *t* tests: **pooled variances** and **separate variances**. The former is used when the two population variances are similar to one another, while the latter is for use when the variances are significantly disparate.

The mental sequence you should use when deciding what kind of *t* test to use is depicted in Figure 11.2.

Pooled variances: The type of *t* test appropriate when the samples are independent and the population variances are equal.

Separate variances: The type of *t* test appropriate when the samples are independent and the population variances are unequal.

In this chapter, we will encounter something we have seen before but have not addressed in detail: **one-tailed** versus **two-tailed** tests. The *t* distribution, as you know, is symmetrical and has positive

Figure 11.2   Steps for Deciding Which *t* Test to Use

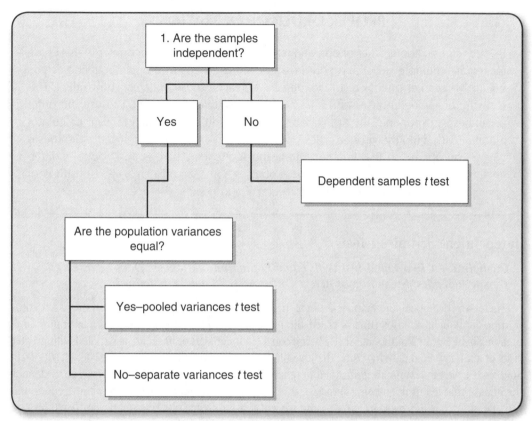

and negative sides. We touched on this topic in Chapter 8. In two-tailed tests, there are two critical values, one positive and one negative. You learned in Chapter 8 that confidence intervals are always two-tailed. In *t* tests, by contrast, some analyses will be two-tailed and some will be one-tailed. Two-tailed tests split the α level such that half of the area represented by alpha is in each tail of the *t* distribution; one-tailed tests place all of α into a single tail. One-tailed tests may have α in the upper (or positive) tail or lower (or negative) tail, depending on the specific question under investigation. Let us work our way through some examples and discuss one-tailed and two-tailed tests as we go.

One-tailed tests: Hypothesis tests in which the entire alpha is placed in either the upper (positive) or lower (negative) tail such that there is only one critical value of the test statistic. Also called *directional tests*.

Two-tailed tests: Hypothesis tests in which alpha is split in half and placed in both tails of the distribution such that there are two values of the test statistic. Also called *nondirectional tests*.

STUDY TIP: CALCULATORS AND PROPER ORDER OF OPERATIONS

Be very careful about order of operations! The formulas we will encounter in this chapter require multiple steps, and you have to do those steps in proper sequence or you will arrive at an erroneous result. Remember "*Please Excuse My Dear Aunt Sally*"? This is a silly but nonetheless useful mnemonic device that reminds you to use the order parentheses, exponents, multiplication, division, addition, subtraction. Your calculator automatically employs proper order of operations, so you need to insert parentheses where appropriate so that you can direct the sequence. There is an enormous difference between, for instance, the numeric phrases *3 + 4/2* and *(3 + 4)/2*. Consult your operator's manual and your course instructor if you need assistance.

Independent Samples *t* Tests

Example 1—A Two-Tailed Test With Equal Population Variances: Does Transferred Female Juveniles' Mean Age of Arrest Differ Across Races/Ethnicities?

There is a theoretical and empirical connection between how old people are when they start committing delinquent offenses (this is called the *age of onset*) and their likelihood of continuing lawbreaking behavior in adulthood. All else being equal, younger ages of onset are associated with greater risks of adult criminal activity. Using the Juvenile Defendants in Criminal Court data set (JDCC; Data Sources 10), we can test for a significant difference in the mean age at which juveniles were arrested for the offense that led to them being transferred to adult court. To address questions about gender and race, the sample is narrowed to females, and we will test for an age difference between Hispanics and whites in this subsample. Among Hispanic female juveniles in the JDCC sample ($N = 44$), the mean age of arrest was 15.89 years ($s = 1.45$). Among white females ($N = 31$), the mean age of arrest was 16.57 years ($s = 1.11$). To be clear, the IV is race (*Hispanic; white*) and the DV is age at arrest (*years*). A *t* test is therefore the proper analytic strategy. We will use an α level of .05, a presumption that the population variances are equal, and the five steps of hypothesis testing.

It is useful in an independent samples *t* test to first make a table that lists the relevant pieces of information that you will need for the test. Table 11.1 shows these numbers. It does not matter which sample you designate Sample 1 and which you call Sample 2 as long as you stick with your original designation throughout the course of the hypothesis test. Since it is easy to simply designate the samples in the order in which they appear in the problem, let us call Hispanic females Sample 1 and white females Sample 2.

Table 11.1   Relevant Numbers for Example 1

Sample 1: Hispanic Females	Sample 2: White Females	Test
$\bar{x}_1 = 15.89$	$\bar{x}_1 = 16.57$	$\alpha = .05$
$s_1 = 1.45$	$s_1 = 1.11$	two-tailed test
$N_1 = 44$	$N_1 = 31$	equal population variances

This is a two-tailed test because the alternative hypothesis merely specifies a difference (that is, an inequality) between the means—no prediction is being made about which mean is greater than or less than the other one. For this reason, two-tailed tests are also referred to as nondirectional tests. Two-tailed or nondirectional hypothesis tests are used when researchers do not wish to make a prediction that specifies one mean as being the larger or the smaller one. In this example, we are not predicting that Hispanic females' are either significantly younger or significantly older than white females; we have merely posited that the mean ages differ by race/ethnicity.

Step 1: State the Null and Alternative Hypotheses

In *t* tests, the null and alternative are phrased in terms of the population means. Recall that population means are symbolized μ (the Greek letter mu, pronounced "mew"). We use the population symbols rather than the sample symbols because the goal is to make a statement about the relationship, or lack thereof, between two variables in the population. The null hypothesis for a *t* test is that the means are equal:

$$H_0: \mu_1 = \mu_2.$$

Equivalence in the DV means suggests that the IV is not exerting an impact on this outcome measure. Another way of thinking about this is that H_0 predicts that the two samples actually came from the same population. In the context of the present example, retaining the null would indicate that race or ethnicity does not affect female juveniles' age of arrest and that all female juveniles are part of the same population, irrespective of race or ethnicity.

The alternative hypothesis is that there *is* a significant difference between the population means or, in other words, that there are two separate populations, each with its own mean:

$$H_1: \mu_1 \neq \mu_2.$$

Rejecting the null would lead to the conclusion that the IV does affect the DV; here, it would mean that race and ethnicity does appear related to age of arrest.

Step 2: Identify the Distribution and Compute the Degrees of Freedom

As mentioned earlier, two-population tests for differences between means employ the *t* distribution (hence their colloquial name, *t* tests). The *t* distribution, you should recall, is a family of curves that changes shape depending on degrees of freedom (*df*). The *df* formula differs across the three types of *t* tests, so you have to identify the proper test before you can compute the *df*. Using the sequence depicted in Figure 11.2, we know (1) that the samples are independent because this is a random sample divided into two groups; and (2) that the population variances are equal. This leads us to choose the pooled variances *t* test. The *df* formula is

$$df = N_1 + N_2 - 2, \text{ where} \qquad \boxed{Formula\ 11(1)}$$

N_1 = the size of the first sample,

N_2 = the size of the second sample.

Pulling the sample sizes from Table 11.1, the *df* are

$$df = 44 + 31 - 2 = 73.$$

Step 3: Identify the Critical Value and State the Decision Rule

Three pieces of information are required to find the critical value of *t* (t_{crit}) using the *t* table: the number of tails in the test, the alpha level, and the *df*. With two tails, an α of .05, and 73 degrees of freedom, the absolute value of t_{crit} is 2.000. Remember that when the exact *df* you are looking for is not on the table, you should use the number that is closest to yours. Here, the closest *df* to 73 is 60.

This is not the end of finding the critical value, though, because we still have to figure out the sign of each critical value; that is, we need to determine whether t_{crit} is positive, negative, or both. In two-tailed tests, there are always two critical values. Their absolute values are the same, but one is negative and one is positive. Figure 11.3 illustrates this.

Given that there are two tails in this current test and, therefore, two critical values, $t_{crit} = \pm 2.000$. The decision rule is stated thus: *If t_{obt} is either greater than 2.000 or less than −2.000, H_0 will be rejected.* If, by contrast, t_{obt} is less than 2.000 or greater than −2.000, the null will be retained.

Step 4: Calculate the Obtained Value of the Test Statistic

The formulas for the obtained value of *t* (t_{obt}) vary across the three different types of *t* tests; however, the common thread is to have (1) a measure of the difference between means in the numerator and (2) an estimate of the standard error of the sampling distribution of differences between means in the denominator (the standard error is the standard deviation of a sampling

Figure 11.3   The Critical Values for a Two-Tailed Test With $\alpha = .05$ and $df = 73$

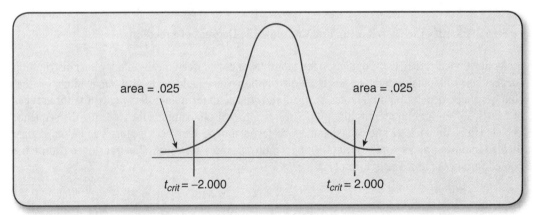

distribution). The standard error is symbolized $\hat{\sigma}_{\bar{x}_1-\bar{x}_2}$ and the formula for estimating it with pooled variances is

$$\hat{\sigma}_{\bar{x}_1-\bar{x}_2} = \sqrt{\frac{(N_1-1)s_1^2 + (N_2-1)s_2^2}{N_1+N_2-2}} \sqrt{\frac{N_1+N_2}{N_1 N_2}}.$$

<figure>Formula 11(2)</figure>

This formula may look a bit daunting, but keep in mind that it is only composed of sample sizes and standard deviations, both of which are numbers you are accustomed to working with. The most important thing is to work through it carefully. Plug the numbers in correctly and use proper equation-solving techniques. Entering the numbers from our example yields:

$$\hat{\sigma}_{\bar{x}_1-\bar{x}_2} = \sqrt{\frac{(44-1)1.45^2 + (31-1)1.11^2}{44+31-2}} \sqrt{\frac{44+31}{44 \bullet 31}} = \sqrt{\frac{(43)2.103 + (30)1.23}{73}} \sqrt{\frac{75}{1364}} = \sqrt{\frac{90.43+36.90}{73}} \sqrt{.05}$$

$$= \sqrt{1.74}\sqrt{.05} = 1.32 \bullet .22 = .29.$$

Now we plug $\hat{\sigma}_{\bar{x}_1-\bar{x}_2}$ into the t_{obt} formula, which is

$$t_{obt} = \frac{\bar{x}_1 - \bar{x}_2}{\hat{\sigma}_{\bar{x}_1-\bar{x}_2}}.$$

<figure>Formula 11(3)</figure>

Using our numbers, we perform the calculation:

$$t_{obt} = \frac{15.89-16.57}{.29} = \frac{-.68}{.29} = -2.34.$$

Therefore, $t_{obt} = -2.34$. Step 4 is done.

Step 5: Make a Decision About the Null and State the Substantive Conclusion

We said in the decision rule that if t_{obt} was either greater than 2.000 or less than −2.000, the null would be rejected. So, what will we do? If you said "reject the null," you are correct. Since t_{obt} is less than −2.000, the null is rejected. The conclusion is that white and Hispanic female juveniles transferred to adult court differ significantly in terms of mean age of arrest for their instant offense.

Another way to think about it is that there is a relationship between race/ethnicity and mean age of arrest.

Example 2—A One-Tailed Test With Unequal Population Variances: Does the length of an incarceration sentence differ depending on whether it was imposed pursuant to a jury trial or a bench trial?

For the second t test example, we will consider whether the length of prison terms imposed upon defendants after conviction differs depending on whether the conviction was handed down by a jury or by a judge. It will be hypothesized that persons convicted by juries are sentenced to significantly longer prison terms relative to those defendants who are convicted via bench trial. We will use the National Judicial Reporting Program data (NJRP; Data Sources 3.1). The sample is narrowed to persons convicted of weapons offenses, as these are serious offenses that may evoke an especially severe response from jurors and judges alike. Weapons offenders convicted by juries ($N = 63$) were sentenced to a mean of 33.97 months in prison ($s = 48.75$), while those convicted by judges ($N = 94$) saw an incarceration mean of 17.26 months ($s = 14.47$). Using an alpha level of .01 and the assumption that the population variances are unequal, we will conduct a five-step hypothesis test to determine whether persons convicted by juries receive significantly longer prison sentences. Table 11.2 shows the numbers we will need for the analysis.

Table 11.2   Relevant Numbers for Example 2

Sample 1: Jury Conviction	Sample 2: Judge Conviction	Test
$\bar{x}_1 = 33.97$	$\bar{x}_1 = 17.26$	$\alpha = .01$
$s_1 = 48.75$	$s_1 = 14.47$	one-tailed test
$N_1 = 63$	$N_1 = 94$	unequal population variances

Step 1: State the Null and Alternative Hypotheses

The null hypothesis is the same as that used above ($H_0: \mu_1 = \mu_2$) and reflects the prediction that the two means do not differ. The alternative hypothesis used in Example 1, however, does not apply in the present context because this time, we are making a prediction about which mean will be greater than the other. The nondirectional $\neq$ sign must therefore be replaced by a sign that indicates a specific direction. This will either be a greater than ($>$) or less than ($<$) sign. We are predicting that jury trials will result in significantly longer mean sentences, so we can conceptualize the hypothesis as *jury trial sentences>bench trial sentences*. Since offenders tried by jury will be Sample 1 and those by bench Sample 2 based on their ordering in the problem, the alternative hypothesis is

$$H_1: \mu_1 > \mu_2.$$

Step 2: Identify the Distribution and Compute the Degrees of Freedom

The distribution is still t, but the df equation for unequal population variances differs sharply from that for equal variances because unequal variances mandates the use of the separate variances t test.

The *df* formula is obnoxious, but as with the prior formulas we have encountered, you have everything you need to solve it correctly—just take care to plug in the right numbers and use proper order of operations.

$$df = \left[\frac{\left(\dfrac{s_1^2}{N_1 - 1} + \dfrac{s_2^2}{N_2 - 1} \right)^2}{\left(\dfrac{s_1^2}{N_1 - 1} \right)^2 \left(\dfrac{1}{N_1 + 1} \right) + \left(\dfrac{s_2^2}{N_2 - 1} \right)^2 \left(\dfrac{1}{N_2 + 1} \right)} \right] - 2.$$

Formula 11(4)

Plugging in the correct numbers from the current example,

$$df = \left[\frac{\left(\dfrac{48.75^2}{63 - 1} + \dfrac{14.47^2}{94 - 1} \right)^2}{\left(\dfrac{48.75^2}{63 - 1} \right)^2 \left(\dfrac{1}{63 + 1} \right) + \left(\dfrac{14.47^2}{94 - 1} \right)^2 \left(\dfrac{1}{94 + 1} \right)} \right] - 2 = \left[\frac{\left(\dfrac{2376.56}{62} + \dfrac{209.38}{93} \right)^2}{\left(\dfrac{2376.56}{62} \right)^2 \left(\dfrac{1}{64} \right) + \left(\dfrac{209.38}{93} \right)^2 \left(\dfrac{1}{95} \right)} \right] - 2$$

$$= \left[\frac{(38.33 + 2.25)^2}{(38.33)^2 (.02) + (2.25)^2 (.01)} \right] - 2 = \left[\frac{40.58^2}{1469.19(.02) + 5.06(.01)} \right] - 2 = \left[\frac{1646.74}{29.38 + .05} \right] - 2$$

$$= \left[\frac{1646.74}{29.43} \right] - 2 = 55.95 - 2 = 53.95.$$

Step 3: Identify the Critical Value and State the Decision Rule

With one tail, an α of .01, and $53.95 \approx 54$ degrees of freedom, $t_{crit} = 2.390$. The sign of the critical value is positive because the alternative hypothesis predicts that $\mu_1 > \mu_2$. Revisit Figure 11.2 for an illustration. When the alternative predicts that $\mu_1 < \mu_2$, the critical value will be on the left (negative) side of the distribution and when the alternative is that $\mu_1 > \mu_2$, t_{crit} will be on the right (positive) side. The decision rule is: *If t_{obt} is greater than 2.390, H_0 will be rejected.*

Step 4: Compute the Obtained Value of the Test Statistic

As before, the first step is to obtain an estimate of the standard error of the sampling distribution. Since the population variances are unequal, the separate variances version of independent samples *t* must be used. The standard error formula for this test is

$$\hat{\sigma}_{\bar{x}_1 - \bar{x}_2} = \sqrt{\frac{s_1^2}{N_1 - 1} + \frac{s_2^2}{N_2 - 1}}.$$

Formula 11(5)

Plugging in the numbers from the present example yields

$$\hat{\sigma}_{\bar{x}_1 - \bar{x}_2} = \sqrt{\frac{48.75^2}{63-1} + \frac{14.47^2}{94-1}} = \sqrt{\frac{2376.56}{62} + \frac{209.38}{93}} = \sqrt{38.33 + 2.25} = \sqrt{40.58} = 6.37.$$

Now, the standard error estimate can be entered into the same t_{obt} formula used with the pooled variances t test. Using Formula 11(3),

$$t_{obt} = \frac{33.97 - 17.26}{6.37} = 2.62.$$

Step 5: Make a Decision About the Null and State the Substantive Conclusion

The decision rule stated that the null would be rejected if t_{obt} exceeded 2.390. Since t_{obt} ended up being 2.62, the null is rejected. We conclude that weapons offenders convicted by juries are given significantly longer mean prison sentences, on average, as compared to their counterparts who are convicted by judges. Another way to state this is that there is a relationship between mode of conviction and prison sentence length, with persons convicted by juries receiving significantly longer sentences than persons convicted by judges.

Dependent Samples *t* Tests

The foregoing discussion centered on the situation in which a researcher is working with two independently selected samples; however, as described above, there are times when the samples under examination are not independent. The main types of dependent samples are *matched pairs* and *repeated measures*. Dependent samples require a *t* formula different from that used when the study samples are independent because of the manipulation entailed in selecting dependent samples. With dependent samples *t*, the sample size (N) is not the total number of people or objects in the sample but, rather, the number of *pairs* being examined. We will go through an example now to demonstrate the use of this *t* test.

Example 3—Dependent Samples t Test: Are Female Correctional Officers a Threat to Institutional Security?

The traditionally male-dominated field of correctional security is gradually being opened to women who wish to work in these environments, yet there are lingering concerns regarding how well female correctional officers can maintain order in male institutions. Criticism has been raised that women are not as capable as men when it comes to controlling male inmates, which could threaten the internal safety and security of the prison environment. Let us test the hypothesis that male maximum-security prisons with relatively small percentages of female security staff will have lower inmate-on-inmate assault rates relative to those institutions with high percentages of female security staff because security in the latter will be compromised. We will use data from the Census of State and Federal Adult Correctional Facilities (CSFACF; Data Sources 7) and an alpha level of .05. The first sample consists of five male, state-run, maximum security prisons in Texas with below-average percentages of female security staff, and the second sample contains five prisons selected on the basis of each one's similarity to a prison in the first sample (in other words, the second sample's prisons are all male, state-run, maximum security facilities in Texas with inmate totals similar to those of the first sample). The difference between the samples is that the second sample has above-average percentages of female security staff. Table 11.3 contains the raw data.

Table 11.3   Matched Samples of Male Maximum Security Prisons in Texas

	Assault Rate	
	Low Percentage Female(x_1)	*High Percentage Female(x_2)*
Pair A	1.46	.63
Pair B	2.72	.76
Pair C	1.22	1.29
Pair D	.95	1.09
Pair E	1.26	1.72

Step 1: State the Null and Alternative Hypotheses

The null, as is always the case with *t* tests, is H_0; $\mu_1 = \mu_2$. It is being suggested in this problem that low percentage female prisons should have lower assault rates than high percentage female prisons do; hence the alternative in words is *low < high*. The alternative is therefore: H_1: $\mu_1 < \mu_2$.

Step 2: Identify the Distribution and Compute the Degrees of Freedom

The distribution is *t* and the *df* for dependent samples is

$$df = N_{pairs} - 1.$$ ◁ *Formula 11(6)*

Using the data from Example 3:

$$df = 5 - 1 = 4.$$

Step 3: Identify the Critical Value and State the Decision Rule

With an alpha of .05, a one-tailed test, and 4 degrees of freedom, the absolute value derived using the *t* table is 2.132. To determine whether this critical value is positive or negative, refer back to Figure 11.1. The phrasing of our alternative hypothesis makes the critical value negative, so $t_{crit} = -2.132$. The decision rule states that *if $t_{obt} < -2.132$, H_0 will be rejected.*

Step 4: Compute the Obtained Value of the Test Statistic

The formulas required to calculate t_{obt} for dependent samples looks quite different from those for independent samples, but the logic is the same. The numerator contains a measure of the difference between means, and the denominator is an estimate of the standard error. Finding the standard error requires two steps. First, the standard deviation of the differences between means (s_D) is computed using the formula

$$s_D = \sqrt{\frac{\Sigma\left(x_D - \bar{x}_D\right)^2}{N_{pairs} - 1}}, \text{ where}$$

Formula 11(7)

x_D = the difference scores,

$\bar{x}_D$ = the mean of the difference scores.

The standard deviation is then used to find the standard error of the sampling distribution, as follows:

$$\hat{\sigma}_{\bar{x}_1 - \bar{x}_2} = \frac{s_D}{\sqrt{N_{pairs}}}.$$

Formula 11(8)

Finally, the obtained value of the test statistic is calculated as

$$t_{obt} = \frac{\bar{x}_D}{\hat{\sigma}_{\bar{x}_1 - \bar{x}_2}}.$$

Formula 11(9)

This computation process is substantially simplified by the use of Table 11.4. This table contains the raw scores and three additional columns. The first column to the right of the raw scores contains the difference scores (x_D), which are computed by subtracting each x_2 score from its corresponding x_1 value. The mean difference score ($\bar{x}_D$) is then needed and is calculated thus:

$$\bar{x}_D = \frac{\Sigma x_D}{N}.$$

Formula 11(10)

Table 11.4 Matched Samples of Male Maximum Security Prisons in Texas

| | Assault Rate by Percent Female | | | | |
	Low Percent Female (x_1)	High Percent Female (x_2)	$x_D = (x_1 - x_2)$	$(x_D - \bar{x}_D)$	$(x_D - \bar{x}_D)^2$
Pair A	1.46	.63	.63 – 1.46 = .83	.83 – .42 = .41	$(.41)^2 = .17$
Pair B	2.72	.76	1.96	1.54	2.37
Pair C	1.22	1.29	−.07	−.49	.24
Pair D	.95	1.09	−.14	−.56	.31
Pair E	1.26	1.72	−.46	−.88	.77
			$\bar{x}_D = \dfrac{2.12}{5} = .42$		$\Sigma = 3.86$

The mean difference score is subtracted from each individual difference score—this is the $(x_D - \bar{x}_D)$ column—to form deviation scores. Finally, each of these scores is squared and the last column summed. The final product of Table 11.4 is the sum of the squared deviation scores, located in the lower right-hand corner. This number (here, 3.86) gets entered into the s_D formula and the calculations proceed through t_{obt} as such:

$$s_D = \sqrt{\frac{3.86}{5-1}} = \sqrt{.97} = .98$$

$$\sigma_{\bar{x}_1 - \bar{x}_2} = \frac{.98}{\sqrt{5}} = \frac{.98}{2.24} = .44$$

$$t_{obt} = \frac{.42}{.44} = .95.$$

Step 5: Make a Decision About the Null and State the Substantive Conclusion

The decision rule in this problem was that the null would be rejected if the obtained value was less than −2.132. With $t_{obt} = .95$, the null is retained. The conclusion is that male maximum security prisons with low levels of female security staff do not experience fewer inmate-on-inmate assaults relative to prisons with high levels of female security staff. There is no support for the notion that female correctional officers compromise institutional security.

RESEARCH EXAMPLE 11.2

Pulling Levers: Targeted Interventions to Reduce Crime

The last two decades have seen a shift in criminal justice operations toward an emphasis on multiagency efforts designed to address the specific crime problems plaguing individual neighborhoods and communities. One strategy that has been devised is the "pulling levers" approach that entails identifying high-risk offenders and offering them a choice between tough prosecution and reform. Police and prosecutors explain the penalties these offenders would face if they continue their antisocial behavior, while community social service providers offer treatment and educational opportunities for those who want to put their lives on a better track. In 2007, the Rockford, Illinois, Police Department implemented a pulling levers approach in an attempt to reduce violent crime in the city. Corsaro, Brunson, and McGarrell (2009) evaluated the Rockford intervention. They measured the mean number of violent crimes that

(Continued)

(Continued)

occurred in several areas of the city per month before and after the program. The table below shows these results. The target neighborhood is the one in which the pulling levers strategy was implemented.

	Number of Offenses per Month	
	Preintervention	Postintervention
Location		
Target Neighborhood		
Nonviolent	29	22
Violent	21	18
Remainder of City		
Nonviolent	944	859
Violent	567	554
Overall City		
Nonviolent	1013	881
Violent	588	573

Using a statistical analysis, the researchers found that the target neighborhood experienced a significant reduction in nonviolent crime as a result of the pulling levers program. The reduction in violent crime, though, was not statistically significant, so it appeared that the intervention did not affect these types of offenses. The researchers were able to attribute the target area's decline in nonviolent offenses to the pulling levers strategy because crime fell only in the target zone and nowhere else in the city. It thus appears from this analysis that pulling levers is a promising strategy for reducing nonviolent crimes such as drug, property, and nuisance offenses, but is perhaps less useful with respect to violent crime.

▣ TWO-POPULATION TESTS FOR DIFFERENCES BETWEEN PROPORTIONS

Two-population tests for differences between proportions follow the same logic as those for differences between means. The independent variable is still a two-class, categorical measure; however, the dependent variable is a proportion rather than a mean. Differences between proportions have their own sampling distribution, which looks very much like that for differences between means and can be drawn as Figure 11.5.

Figure 11.4   The Sampling Distribution of Differences Between Proportions

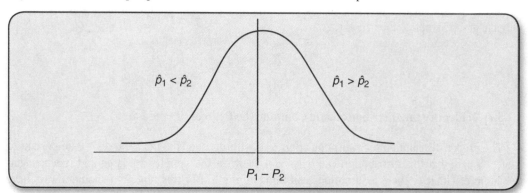

Population proportions are symbolized with the letter P and sample proportions with $\hat{p}$ (pronounced "p hat"). Since you already know the fundamentals behind two-population tests, we will dive right into an example.

Example 4: Do community-based sex offender treatment programs significantly reduce recidivism?

Sex offenders are perhaps the most reviled class of criminal offenders, yet imprisoning low-level sex offenders is not necessarily the best option in terms of social and economic policy. These people might be good candidates for community-based treatment, which can be effective at reducing the likelihood of recidivism and is less expensive than imprisonment. Washington State has a program called the Special Sex Offender Sentencing Alternative (SSOSA) that separates low-risk from high-risk sex offenders and sentences the low-risk offenders to community-based supervision and treatment. Participants must maintain standards of good behavior, compliance with imposed conditions, and adherence to the required treatment regimen. Transgressors are removed from SSOSA and incarcerated. This program has raised questions about how well the SSOSA program reduces recidivism. It has also generated concerns among critics who fear that allowing convicted sex offenders to remain in the community jeopardizes public safety.

To address the issue of sex offender recidivism and the effectiveness of the SSOSA program, researchers from the Washington State Institute for Public Policy compared recidivism across two groups of male sex offenders: The first group was given SSOSA sentences and the second met eligibility requirements for the SSOSA program but were instead sentenced to prison and did not receive treatment during their period of incarceration. Recidivism was measured as a rearrest for a new sex offense within seven years of completion or release. They found that the SSOSA group ($N = 321$) had a recidivism rate of 11% and the SSOSA-eligible group ($N = 306$) recidivated at a rate of 14% (Song & Lieb, 1995). Using an alpha level of .05, let us test the null hypothesis that there is no difference between the population proportions against the alternative hypothesis that the SSOSA group's recidivism rate was significantly lower than the SSOSA-eligible group's.

Step 1: State the Null and Alternative Hypotheses

The null hypothesis for a two-population test for differences between proportions represents the prediction that the two proportions do not differ or, in other words, that the two population proportions are equal. The alternative hypothesis can take on the three forms discussed in the preceding section regarding two-population tests for means. Here, calling the SSOSA group Sample 1 and the SSOSA-eligible group Sample 2, the hypotheses are

$$H_0: P_1 = P_2$$

$$H_1: P_1 < P_2.$$

Step 2: Identify the Distribution and Compute the Degrees of Freedom

The proper distribution for two-population tests of proportions is the z curve. It is important to note, though, that this distribution can only be used when the samples are large and independent; violation of either of these assumptions can be fatal to a test of this type. You should always have enough knowledge about the research design that produced the data you are working with to determine whether the independence assumption has been met. The way to determine whether the large sample criterion holds is to ensure that for each of the samples

$$N\hat{p}_k \geq 5$$ Formula 11(11)

and

$$N\hat{q}_k \geq 5, \text{ where}$$

$\hat{p}_k$ = the sample proportion for each of Sample 1 and Sample 2,

$\hat{q}_k = 1 - \hat{p}_k$ for each of Sample 1 and Sample 2.

In the current example, the SSOSA group meets the large sample criterion because $\hat{p}_1 = .11$, which means that $\hat{q}_1 = 1.00 - .11 = .89$. Therefore:

$$321(.11) = 35.31$$

$$321(.89) = 285.69.$$

The SSOSA-eligible group likewise succeeds, as $\hat{p}_2 = .14$, $\hat{q}_2 = 1.00 - .14 = .86$, and

$$306(.14) = 42.84$$

$$306(.86) = 263.16.$$

The z distribution can thus be used. There is no need to compute degrees of freedom because they are not applicable to z.

Step 3: Identify the Critical Value and State the Decision Rule

It has been awhile since we used the z distribution, but recall that a z value can be found using a known area. Here, alpha is that area. Since $\alpha = .05$ and the test is one-tailed, go to the z table and find the area closest to $.500 - .05 = .450$. There are actually two areas that fit this description (.4495 and

.4505), so $z_{crit} = \frac{-1.64 + (-1.65)}{2} = -1.645 \approx -1.65$. The critical value is negative because the alternative hypothesis $(P_1 < P_2)$ tells us that we are working on the left (negative side) of the distribution (see Figure 11.5). The decision rule is thus: *If $z_{obt} < -1.65$, H_0 will be rejected.*

Step 4: Compute the Obtained Value of the Test Statistic

For this portion of the test, you need to be very careful about symbols because there are a few that are similar to one another but actually represent very different numbers. Pay close attention to the following:

P = population proportion

$\hat{p}$ = the pooled sample proportions as an estimate of the population proportion

$\hat{q}$ = 1.00 – the pooled sample proportions

$\hat{p}_1$ = Sample 1 proportion

$\hat{p}_2$ = Sample 2 proportion

There are a few analytical steps in the buildup to the z_{obt} formula. First, the sample proportions have to be pooled in order to form a single estimate of the proposed population proportion:

$$\hat{p} = \frac{N_1 \hat{p}_1 + N_2 \hat{p}_2}{N_1 + N_2}.$$

Formula 11(12)

The complement of the pooled proportion ($\hat{q}$) is also needed. This is done using the formula

$$\hat{q} = 1.00 - \hat{p}.$$

Formula 11(13)

Next, the standard error of the sampling distribution of differences between proportions must be estimated using the formula

$$\hat{\sigma}_{\hat{p}_1 - \hat{p}_2} = \sqrt{\hat{p}\hat{q}}\sqrt{\frac{N_1 + N_2}{N_1 N_2}}.$$

Formula 11(14)

Finally, the obtained value of the test statistic is calculated as

$$z_{obt} = \frac{\hat{p}_1 - \hat{p}_2}{\hat{\sigma}_{\hat{p}_1 - \hat{p}_2}}.$$

Formula 11(15)

Now, we will plug in the numbers from the current example and solve the formulas all the way up to z_{obt}:

$$\hat{p} = \frac{321(.11) + 306(.14)}{321 + 306} = \frac{35.31 + 42.84}{627} = \frac{78.15}{627} = .12$$

$$\hat{q} = 1 - .12 = .88$$

$$\sigma_{\hat{p}_1 - \hat{p}_2} = \sqrt{(.12)(.88)} \sqrt{\frac{321 + 306}{(321)(306)}} = \sqrt{.11} \sqrt{\frac{627}{98226}} = \sqrt{.11} \sqrt{.01} = (.33)(.10) = .03$$

$$z_{obt} = \frac{.11 - .14}{.03} = \frac{-.03}{.03} = -1.00.$$

Step 5: Make a Decision About the Null and State the Substantive Conclusion

$Z_{obt} = -1.00$, which is not less than -1.65, so the null is retained. The SSOSA program does not appear to significantly reduce participants' commission of new sex offenses; however, SSOSA participants are no *more* dangerous than are their imprisoned counterparts. This community-based program, then, does not appear to either help or harm public safety.

▣ SPSS

The SPSS program can run all of the t tests discussed in this chapter, though it cannot run tests for differences between proportions, so you would have to use SPSS to derive the proportions and then do the analysis by hand. Independent samples t tests are located under the *Analyze* menu in the SPSS data screen. In this menu, find *Compare Means* and then *Independent-Samples T Test*. To demonstrate the use of SPSS, some of the examples we did by hand in the foregoing pages will be replicated. Note that the final answers obtained in SPSS may depart somewhat from those that we calculated by hand because we use only two decimal places and SPSS uses far more than that.

First, let us run an independent-samples t test using the *trial type* and *prison sentence length* variables that were under study in Example 2 above. Figure 11.5 shows how to select these variables for an analysis.

The IV is designated as the *Grouping Variable* and the DV goes into the *Test Variable(s)* space. Click *OK* to obtain the output shown in Figure 11.6.

The first box in Figure 11.6 contains the t test results. Notice that SPSS produces both pooled variances (equal variances assumed) and separate variances (equal variances not assumed) tests. You have to decide which set of results is the correct one; you do this using the Levene's F statistic. Levene's F is a test of similarity between the variances. The null hypothesis is that they are equal. It is customary to use an alpha level of .05 for assessing the validity of the null. If the p value for Levene's F (recall that p values are found in *Sig.* columns in output) is less than .05, the null of equality is rejected and you should use the separate variances (that is, equal variances not assumed) t value. If $p < .05$, the null is retained and you should use the output for the pooled/equal variances test.

You can see in Figure 11.6 that Levene's F (5.778) is statistically significant because $.017 < .05$; therefore, the null of equality is rejected and the proper test is separate variances t. The obtained value in the output is 2.644, which is very similar to our hand-calculated value of 2.62. The p value for t_{obt} is .01, which is exactly equal to the stated α level of .01. When this happens, the null is generally rejected; however, it is wise to be humble when you are in this situation and to say that the null has been rejected at an α of .05 rather than of .01.

Figure 11.5   Selecting Variables for Two-Population Tests

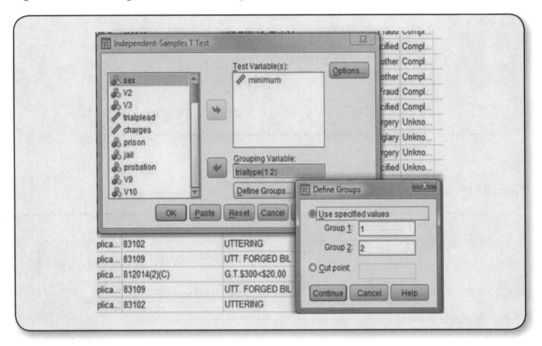

Figure 11.6   Independent Samples *t* Test Output

		Levene's Test for Equality of Variances		t-test for Equality of Means						
									95% Confidence Interval of the Difference	
		F	Sig.	t	df	Sig. (2-tailed)	Mean Difference	Std. Error Difference	Lower	Upper
Minimum incarceration term	Equal variances assumed	5.778	.017	3.129	155	.002	16.71293	5.34118	6.16204	27.26383
	Equal variances not assumed			2.644	69.378	.010	16.71293	6.32010	4.10590	29.31997

Running dependent samples *t* tests involves a similar procedure, though the command menu for dependent samples differs from that for independent samples. As shown in Figure 11.7, rather than choosing your independent and dependent variables, you must enter the raw scores into two columns (one column for Sample 1's scores and one column for Sample 2's) and then select them for analysis. Figure 11.8 shows the output for the analysis using the two matched samples of male, maximum-security prisons in Texas that were used in Example 3.

The obtained value of *t* is .966, which differs only slightly from the value of .95 that our hand computations yielded. The *p* value is .389. This is well above the alpha level of .05, so the null is retained.

Figure 11.7   Running a Dependent Samples *t* Test

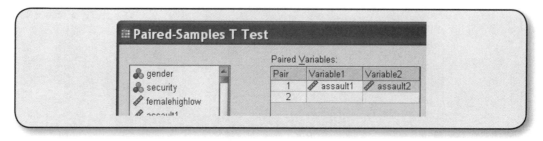

Figure 11.8   Dependent Samples *t* Test Output

Paired Samples Test

		Paired Differences					t	df	Sig. (2-tailed)
					95% Confidence Interval of the Difference				
		Mean	Std. Deviation	Std. Error Mean	Lower	Upper			
Pair 1	assault1 - assault2	.42472	.98292	.43958	-.79573	1.64518	.966	4	.389

CHAPTER SUMMARY

In this chapter, you learned several types of analyses that can be conducted to test for differences between two populations. These can be used to test for differences between means or for differences between proportions; each type of difference has its own sampling distribution. The *t* distribution can be used for means tests and the *z* for proportions tests.

When you approach a hypothesis-testing question involving a two-class, categorical independent variable and a continuous dependent variable, you first have to ask yourself whether the two samples are independent. If they are not, the dependent samples *t* test must be used. If they are independent, you must then decide whether the population variances are equal. If the DV at issue is measured as a proportion, then the two-population tests for differences in proportions should be used.

SPSS can be used to run *t* tests. You should use your knowledge of the research design to determine whether the samples are independent or dependent. If they are independent, Levene's *F* statistic is used to determine whether or not the population variances can be assumed equal. If *F* is not statistically significant (generally at an alpha of .05), then the null of equality is retained and you should use the equal/pooled variances *t*; if the null is rejected at .05, the unequal/separate variances test is the one to look at. Tests for differences in proportions must be done by hand.

As always, GIGO! It is your responsibility to make sure that you select the proper test and run that test correctly. SPSS is pretty unforgiving—it will not alert you to errors unless they are so serious that the requested analysis simply cannot be run at all. If you make a mistake such as using dependent samples when you should use independent or using pooled variances instead of separate, SPSS will give you a result that resembles the serial killers that opened this chapter: It looks normal on the surface, but it is actually untrustworthy and potentially dangerous.

CHAPTER 11 REVIEW PROBLEMS

1. A researcher wants to test the hypothesis that defendants who plead guilty are sentenced more leniently than those who insist on going to trial. The researcher measures the plea decision as *guilty plea; trial* and sentence as the number of months of incarceration to which defendants were sentenced. Answer the following questions.

   a. What is the independent variable?
   b. What is the level of measurement of the independent variable?
   c. What is the dependent variable?
   d. What is the level of measurement of the dependent variable?

2. A researcher wishes to find out whether a new drug court program appears to be effective at reducing drug use among participants. He gathers a random sample of drug defendants who are about to enter the drug court's treatment regimen and measures the number of times per month that they use drugs. After the participants finish the program, the researcher again measures their monthly drug use. Which type of *t* test would be appropriate for analyzing the data?

   a. Independent samples, pooled variances
   b. Independent samples, separate variances
   c. Dependent samples

3. A researcher is investigating the relationship between the restrictiveness of gun laws and gun crime rates. She gathers a sample of states and divides them according into two groups: *strict gun laws; lax gun laws*. She then collects data on the gun crime rate in each state. She finds that the two groups have unequal variances. Which type of *t* test would be appropriate for analyzing the data?

   a. Independent samples, pooled variances
   b. Independent samples, separate variances
   c. Dependent samples

4. A researcher wishes to test the hypothesis that attorney type affects the length of time it takes for a criminal case to be disposed of. He gathers a sample of defendants and records attorney type (*publicly funded; privately retained*) and the number of days from the time charges were filed to the final disposition of the case. He finds that the groups have equal variances. Which type of *t* test would be appropriate for analyzing the data?

   a. Independent samples, pooled variances
   b. Independent samples, separate variances
   c. Dependent samples

5. In Research Example 10.1, you learned of a study by Wright et al. (2008) in which the researchers set out to determine whether multiple homicide offenders (MHOs) were diverse in the number and types of crimes they commit or whether, instead, they tend to specialize in killing. The researchers compared MHOs to single homicide offenders (SHOs) for purposes of this study. MHOs ($N = 155$) had a mean diversity index score of .36 ($s = .32$) and SHOs ($N = 463$) had a mean of .37 ($s = .33$). Using an alpha level of .05, test the null hypothesis of no difference between the means against the alternative hypothesis that there is a statistically significant difference between the mean diversity index scores for MHOs and SHOs. Assume equal population variances. Use all five steps.

6. One persistent area of neglect in policing research is the study of police in nonurban areas. Use the Law Enforcement Management and Administrative Statistics (LEMAS) data from the state of Washington to

Note: Answers to review problems in this and subsequent chapters may vary depending on the number of steps used and whether or not rounding is employed during the calculations. The answers provided in this book's key were derived using the procedures illustrated in the main text.

test whether police levels (measured as police officers per 1,000 citizens) vary across urban and nonurban areas. The LEMAS survey shows that urban jurisdictions in Washington ($N = 13$) have a mean of .84 police officers per 1,000 citizens ($s = .68$) and that nonurban jurisdictions ($N = 40$) have a mean of 1.68 ($s = 1.37$). Using an alpha level of .05, test the null hypothesis that there is no difference between the means against the alternative that they are significantly different. Assume equal population variances. Use all five steps.

7. Continuing with the urban-versus-nonurban theme and the LEMAS data, explore the question now of whether there is a difference between urban and rural jurisdictions in the hiring of female police officers. The dependent variable is *percentage female* and measures the percentage of the entire sworn force in a police agency that is female. The sample is narrowed to California. In urban jurisdictions ($N = 77$), the mean percentage female was 9.65 ($s = 4.17$). Nonurban areas ($N = 84$) had a mean of 6.80 ($s = 4.16$). Using an alpha level of .01, test the null hypothesis that there is no difference between the means against the alternative that urban jurisdictions have a significantly higher mean percentage of female officers. Assume equal population variances. Use all five steps.

8. It is well established that there are substantial regional differences with respect to death sentences, with some regions imposing many death sentences and others relatively few or even none on an annual basis. The vast majority of people sentenced to death, though, are either not executed at all or are executed only after the substantial delay of the appeals process. This raises a question as to whether there are regional differences in the annual number of executions that actually take place. According to the Uniform Crime Reports (UCR), Southern states that had the death penalty ($N = 15$) executed an average of 3.00 people per state in 2009 ($s = 6.07$). Non-Southern states that authorized capital punishment ($N = 22$) averaged .32 executions per state ($s = 1.09$). Using an alpha level of .05, test the null hypothesis of no difference against the alternative that Southern states' mean is significantly higher than that of non-Southern states. Assume unequal population variances. Use all five steps.

9. There are conflicting perspectives regarding whether juveniles transferred to adult courts are treated more leniently due to their young age or whether they experience harsher treatment owing to judges' perceptions that juveniles who get transferred are especially dangerous. The Juvenile Defendants in Criminal Courts (JDCC) data set contains information about juveniles' age at arrest and the sentence length received by those sent to jail pursuant to conviction. Those male juveniles who were under 16 at the time of arrest ($N = 85$) received a mean of 68.84 days in jail ($s = 125.59$) and those who were over 16 at arrest ($N = 741$) had a jail sentence mean of 95.24 days ($s = 146.91$). Using an alpha level of .01, test the null hypothesis of no difference against the alternative hypothesis that the means are significantly different. Assume unequal population variances. Use all five steps.

10. One of the most obvious potential contributors to the problem of assaults against police officers is exposure—all else being equal, jurisdictions wherein officers make more arrests may have elevated rates of officer assaults relative to lower-arrest jurisdictions. The UCR offer state-level information on arrest rates and officer assault rates. The states in the two samples in the table below were selected based upon key similarities; that is, they are all in the Western region of the country, have similar statewide violent crime rates, and have similar populations. The difference is that the states in the first sample have relatively low arrest rates and those in the second sample have relatively high arrest rates. The following table shows the 2009 officer assault rate (number of officers assaulted per 1,000 officers) in each pair of states. Using an alpha level of .05, test the null hypothesis that there is no difference in the group means against the alternative hypothesis that there is a significant difference in officer assaults across the two groups. Use all five steps.

State Pair	Officer Assaults per 1,000 Officers	
	Low Arrest Rate	*High Arrest Rate*
Pair A	2.54	2.28
Pair B	4.72	4.08
Pair C	2.75	2.28
Pair D	2.55	2.13
Pair E	3.31	3.38

11. Capital punishment advocates often assert that the death penalty is a deterrent to homicide because, so the argument goes, would-be killers will think twice before taking a life if they know that they might lose their own as a result. In 2004, the high court in the state of New York declared the state's death penalty legislation unconstitutional. This effectively ended the use of this punishment in New York and provides a good "natural experiment" for testing the predicted deterrent effects of capital punishment. The table below contains UCR homicide data for New York's six largest cities. The years are 2004 and 2005. (Data from 2003 might have been preferable but are not available on a per-city basis.) Using an alpha level of .01, test the null hypothesis of no difference against the alternative that homicide significantly increased in these cities after the death penalty was abolished. Use all five steps.

City	Homicides per 100,000 Citizens	
	2004	*2005*
Amherst Town	0.00	.90
Buffalo	17.86	19.77
New York	7.04	6.64
Rochester	16.70	24.91
Syracuse	11.09	13.26
Yonkers	7.58	4.56

12. Southern states are known for their relatively high levels of gun ownership, which may contribute to higher homicide levels and higher rates of gun involvement in homicide. The 2009 Uniform Crime Reports (UCR) offer state-by-state information about murder weapons, and the states can be divided into *not South* and *South*. The dependent variable is *proportion firearm* and measures the proportion of the murders in each state that were committed with guns. In Southern states ($N = 15$; Florida did not contribute data), .67 of murders in 2009 were committed with firearms. In non-Southern states ($N = 34$), the proportion was .56. Using an alpha of .01, test the null hypothesis of no difference between the proportions against the alternative that Southern states have a significantly higher proportion of firearm-involved murders. Use all five steps.

13. The Sixth Amendment to the U.S. Constitution grants criminal defendants the right to speedy case dispositions, and the American Bar Association (ABA) recommends that defendants who obtain pretrial release should have their cases disposed of within 180 days of their first court appearance. One question with regard to the speed of case processing is whether attorney type matters. Some evidence suggests that publicly appointed attorneys move cases faster than their privately retained counterparts do, other evidence points toward the opposite conclusion, and some studies find no difference. The Juvenile Defendants in Criminal Court (JDCC) data set contains information on attorney type, pretrial release, and days to adjudication. The sample consists of juveniles charged with drug felonies who were granted preadjudication release. Among those juveniles represented by public attorneys ($N = 509$), .64 had their cases disposed of in 180 days or less, while .32 of the juveniles who retained private attorneys ($N = 73$) were adjudicated within 180 days. Using an alpha level of .05, test the null hypothesis of no difference against the alternative that there is a significant difference between the proportions. Use all five steps.

14. Corrections has traditionally been a male-dominated field, but the proportion of prison security staff that is female has been steadily rising over the past few decades. One question of interest is whether there is a difference between private and public prisons in terms of the percentage of security staff that is female. The data set *Census of State and Federal Correctional Facilities for Chapter 11.sav* (**http://www.sagepub .com/gau**) contains the variables *facilitytype* and *pctfemale*. The first variable measures whether a facility is private or public and the second variable is the percentage of security staff that is female. Run a *t* test and answer the following questions.

    a. At an alpha level of .05, will you use the results for the pooled/equal variances *t* or that for separate/unequal variances? How did you make this decision?
    b. What is the obtained value of *t*?
    c. Would you reject the null at an alpha level of .01? Why or why not?
    d. What is your substantive conclusion regarding facility type and female security staff?

15. Is there a significant difference in the number of police officers at the scene of a traffic stop depending on whether the stop takes place during the day or at night? The companion website (**http://www.sagepub .com/gau**) contains a data file called *Police-Public Contact Survey for Chapter 11.sav*. The sample of respondents has been matched according to gender, race, and age. The difference between the groups is that some of the respondents were stopped at night and others during the day. Use SPSS to run a *t* test and answer the following questions.

    a. What is the obtained value of *t*?
    b. Would you reject the null at an alpha level of .05? Why or why not?
    c. What is your substantive conclusion regarding whether or not the number of officers at the scene differs significantly depending on whether the stop takes place during the day or at night?

KEY TERMS

t test	Dependent samples	One-tailed tests
Classes	Pooled variances	Two-tailed tests
Independent samples	Separate variances	

GLOSSARY OF SYMBOLS INTRODUCED IN THIS CHAPTER

t_{crit}	The critical value of t
t_{obt}	The obtained value of t
$\sigma_{\bar{x}_1 - \bar{x}_2}$	The standard error of the sampling distribution of differences between means
s_D	The standard deviation of the differences; used with dependent-samples t
x_D	The raw difference scores; used with dependent-samples t
$\bar{x}_D$	The mean of the difference scores; used with dependent-samples t
p	The population proportion; used with tests for differences between proportions
$\hat{p}$	The sample proportions pooled as an estimate of the population proportion; used with tests for differences between proportions
$\hat{q}$	The complement of $\hat{p}$; used with tests for differences between proportions
$\hat{p}_1$	Sample 1 proportion; used with tests for differences between proportions
$\hat{p}_2$	Sample 2 proportion; used with tests for differences between proportions

REFERENCES

Corsaro, N., Brunson, R. K., & McGarrell, E. F. (2009). Problem-oriented policing and open-air drug markets: Examining the Rockford pulling levers deterrence strategy. *Crime & Delinquency.* Prepublished October 14, 2009. DOI: 10.1177/0011128709345955

Song, L. & Lieb, R. (1995). *Washington State sex offenders: Overview of recidivism studies.* Olympia, WA: Washington State Institute for Public Policy.

Wright, K. A., Pratt, T. C., &DeLisi, M. (2008).Examining offending specialization in a sample of male multiple homicide offenders. *Homicide Studies, 12*(4), 381–398.

Hypothesis Testing With Three or More Population Means

Analysis of Variance

I n Chapter 11, you learned how to determine whether a two-class categorical variable exerts an impact on a continuous outcome measure: This is a case in which a two-population *t* test for differences between means is appropriate. In many situations, though, a categorical IV has more than two classes. The proper analytic technique to use when the independent variable (IV) is categorical with three or more classes and the dependent variable (DV) is continuous is **analysis of variance** (ANOVA).

Analysis of Variance (ANOVA): The analytic technique appropriate when an independent variable is categorical with three or more classes and a dependent variable is continuous.

As its name suggests, ANOVA is a hypothesis-testing procedure premised upon variance. Why do we care about variance when testing for differences between means? Consider the hypothetical distributions displayed in Figure 12.1. The distributions have the same mean but markedly disparate variances—one curve is wide and flat, indicating a fair amount of variance, while the other is taller and thinner, indicating less variance.

In any analysis of differences between means, the variance associated with each mean must be accounted for. This is what ANOVA does. It combines means and variances into a single analysis to test for significant differences between means that would indicate the presence of a relationship between an independent and a dependent variable.

You might be wondering why, if we have a categorical IV and a continuous DV, we do not just use a series of *t* tests to find out if one or more of the means are different from the others. **Familywise error**

Figure 12.1 Hypothetical Distributions With the Same Mean and Different Variances

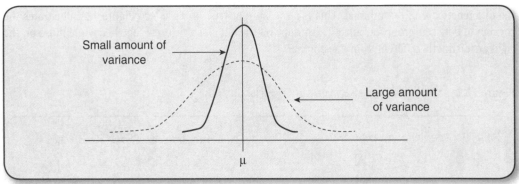

is the reason that this is not a viable analytic strategy. Every time that you run a *t* test, there is a certain probability that the null actually is true (that is, that there is no relationship between the independent and dependent variables) but will be rejected erroneously. This probability is alpha, and the mistake is called a Type I error. Alpha (the probability of incorrectly rejecting a true null) attaches to each *t* test, so in a series of *t* tests, the Type I error rate increases exponentially until the likelihood of mistake reaches an unacceptable level. This is the familywise error rate, and it is the reason that you should never run multiple *t* tests on a single sample.

Familywise error: The increase in the likelihood of a Type I error (erroneous rejection of a true null hypothesis) that results from running repeated statistical tests on a single sample.

Groups: Classes on a categorical independent variable.

Between-group variance: The extent to which a set of groups or classes are similar to or different from one another. This is a measure of true group effect, or a relationship between the independent and dependent variables.

Within-group variance: The amount of diversity that exists among the people or objects in a single group or class. This is a measure of random fluctuation, or error.

▣ ANOVA: DIFFERENT TYPES OF VARIANCES

There are two types of variance analyzed in ANOVA. Both are based on the idea of **groups**, which are the classes on the IV. If an independent variable was *political orientation* measured as *liberal, moderate,* or *conservative*, then liberals would be a group, moderates would be a group, and conservatives would be a group. Groups are central to ANOVA.

The first type of variance is **between-group variance**. This is a measure of the similarity or difference between the groups. It assesses whether groups are quite different from one another or whether the differences are fairly minimal. This is a measure of true group effect. Figure 12.2 illustrates the concept of between-group variance. The groups on the left cluster closely together, while those on the right are distinctly different from one another.

Figure 12.2   Small and Large Between-Group Variability

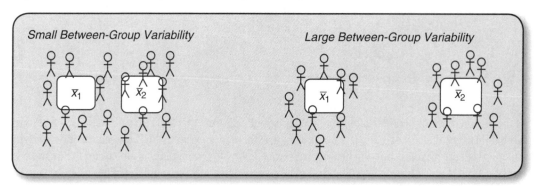

The second kind of variance is **within-group variance** and measures the extent to which people or objects differ from their fellow group members. Within-group variance is driven by random variations between people or objects and is a measure of error. Figure 12.3 depicts the conceptual idea behind within-group variance. The cases in the group on the left cluster tightly around their group's mean, while the cases in the right-hand group are scattered widely around their mean. The left-hand group, then, would be said to have much smaller within-group variability than the right-hand group.

Figure 12.3   Small and Large Within-Group Variability

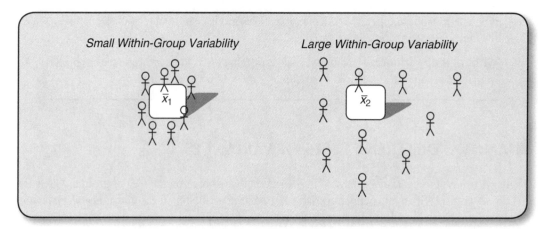

The ANOVA test statistic—called the ***F* statistic** because the theoretical probability distribution for ANOVA is the ***F* distribution**—is a ratio that compares the amount of variance between groups to that within groups. When true differences between groups substantially outweigh the random fluctuations present within each group, the *F* statistic will be large and the null hypothesis that there is no IV-DV relationship will be rejected in favor of the alternative hypothesis that there is an association between the two variables. When between-group variance is small relative to within-group variance, the *F* statistic will be small and the null will be retained.

***F* statistic:** The statistic utilized in ANOVA; a ratio of the amount of between-group variance present in a sample relative to the amount of within-group variance.

***F* distribution:** The sampling/probability distribution for ANOVA. The distribution is bounded at zero on the left and extends to positive infinity; all values in the *F* distribution are thus positive.

Take a moment now to read Research Example 12.1 to see a situation in which criminal justice/criminology researchers would use ANOVA to test for a difference between groups or, in other words, would attempt to discover whether there is a relationship between a multiple-class IV and a continuous DV.

RESEARCH EXAMPLE 12.1

Race-Based Sentencing Disparities: Do Asian defendants benefit from a "model minority" stereotype?

Numerous criminal justice/criminology studies have found racially based sentencing disparities that are not attributable to differences in defendants' prior records or the severity of their instant offenses. Most such studies have focused on white, black, and Hispanic/Latino defendants. One area of the race-and-sentencing research that has received very little scholarly attention is the effect of race on sentencing among Asians. Franklin and Fearn (2010) set out to determine whether Asian defendants are treated differently from those of other races. They predicted that Asians would be sentenced more leniently due to the stereotype in the U.S. that Asians are a "model minority" in that they are widely presumed to be an economically, academically, and socially productive group.

To test the hypothesis that Asian defendants are given lighter sentences relative to similarly-situated defendants of other races, Franklin and Fearn employed the State

(Continued)

(Continued)

Court Processing Statistics data set (SCPS; see Data Sources 12.1). The dependent variable was *sentence length* and was coded as the number of months of incarceration imposed upon offenders sentenced to jail or prison. The authors reported the following statistics with respect to the mean sentence length across race in this sample:

	Defendant Race				
	White	Black	Hispanic	Asian	Total
Mean Sentence Length (Months)	11.80	17.40	16.50	16.10	15.50

Source: Adapted from Table 1 in Franklin and Fearn (2010).

So, what did the researchers find? It turned out that there were no statistically significant differences between the groups. Franklin and Fearn retained the null hypothesis that there is no relationship between race and sentencing, and concluded that Asian defendants do not, in fact, receive significantly shorter jail or prison sentences relative to other racial groups once relevant legal factors (for example, offense type) are taken into account.

Franklin and Fearn's (2010) IV (*race* coded as *white; black; Hispanic; Asian*) was a four-class, categorical variable. Their DV (*sentence length*, measured in months) was continuous. ANOVA is the correct bivariate analysis in this situation. Let us do an example of ANOVA. We will try a spinoff of Franklin and Fearn's analysis using the same data set that they used (the Bureau of Justice Statistics' State Court Processing Statistics; see Data Sources 12) and further examining Asian defendants.

DATA SOURCES 12.1

THE STATE COURT PROCESSING STATISTICS (SCPS) DATA SET

Every two years, the Bureau of Justice Statistics collects data on felony defendants in the nation's most populous counties to form the State Court Processing Statistics data set (Bureau of Justice Statistics, 2006). The SCPS spans from 1990 to 2006 and offers detailed information about defendants and their cases. Sampling is conducted with a two-stage cluster design: First, BJS statisticians randomly select 40 of the 75 most populated counties; and second, each selected county provides data on felony cases filed with the local court on randomly selected days in a single month. Each case is then tracked for up to two years to capture relevant processing information.

Defendant characteristics include age, gender, and race. Case characteristics include variables such as the crime charged, whether adjudication took place via plea or trial, sentencing, and attorney type. This data set is useful to criminal justice/criminology researchers who wish to study felony defendants and/or cases in adult criminal courts.

Example 1: Do Asian defendants' prior criminal records affect the sentences they receive?

We will use the State Court Processing Statistics (SCPS) data to examine whether convicted, incarcerated Asian defendants' prior criminal records influence the length of the prison sentence they receive. For present purposes, the data set is narrowed to the year 2006. The defendants in this sample are Asian males who were convicted of a felony property charge via guilty plea and sentenced to prison. The SCPS data set also contains information on prior convictions, and this variable is coded as *0 prior convictions; 1 – 4 prior convictions; 5+ prior convictions.* This is the independent variable. The dependent variable is sentence length and is measured as the number of years in jail or prison imposed upon the offenders in the sample. Table 12.1 displays the data.

Table 12.1 Maximum Prison Term in Years for Male, Asian Offenders Convicted of Property Crimes, by Criminal History, 2009

	Number of Prior Convictions	
0 Priors (x_1)	*1 – 4 Priors* (x_2)	*5+ Priors* (x_3)
5.00	11.67	1.33
10.00	5.00	2.00
10.00	5.00	1.33
.75	5.00	5.00
$n_1 = 4$	5.00	5.00
	5.00	$n_3 = 5$
	$n_2 = 6$	

We will conduct a five-step hypothesis test to determine whether defendants' conviction histories affect their sentences. Alpha will be set at .01.

Step 1. State the Null and Alternative Hypotheses

The null hypothesis in ANOVA is very similar to that in *t* tests, the only difference being that now there are more than two means. The null is phrased as

$$H_0: \mu_1 = \mu_2 = \mu_3.$$

The structure of the null is dependent upon the number of groups—if there were four groups, there would be a μ_4 as well, and five groups would require the addition of a μ_5.

The alternative hypothesis in ANOVA is a bit different from what we have seen before because the only information offered by this test is whether at least one group is significantly different from at least one other group. The *F* statistic indicates neither the number of differences nor the specific groups or groups that stand out from the others. The alternative hypothesis is, accordingly, rather nondescript. It is phrased as

$$H_1: some\ \mu_i \neq some\ \mu_j.$$

If the null is rejected in an ANOVA test, the only conclusion possible is that at least one group is markedly different from at least one other group—there is no way to tell which group is different or how many between-group differences there are. This is the reason for the existence of **post hoc tests**, which will be covered later in the chapter.

Post hoc tests: Analyses conducted when the null is rejected in ANOVA in order to determine the number and location of differences between groups.

Step 2. Identify the Distribution and Compute the Degrees of Freedom

As aforementioned, ANOVA relies upon the *F* distribution. This distribution is bounded at zero (meaning it has only positive values) and is a family of curves whose shapes are determined by alpha and the degrees of freedom. There are two types of degrees of freedom in ANOVA: between-group degrees of freedom (df_B) and within-groups degrees of freedom (df_W). They are computed as

$$df_B = k - 1$$

Formula 12(1)

$$df_W = N - k, \text{ where}$$

Formula 12(2)

N = the total sample size across all groups,

k = the number of groups.

The total sample size N is derived by summing the number of cases in each group, the latter of which are called *group sample sizes* and are symbolized n_k. In the present example, there are three groups ($k = 3$) and $N = n_1 + n_2 + n_3 = 4 + 6 + 5 = 15$. The degrees of freedom are therefore

$$df_B = 3 - 1 = 2$$

$$df_W = 15 - 3 = 12.$$

Step 3. Identify the Critical Value and State the Decision Rule

The *F* distribution is located in Appendix E. There are different distributions for different alpha levels, so take care to ensure that you are looking at the correct one! You will find the between-group *df* across the top of the table and the within-group *df* down the side. The critical value is located at the intersection of the proper column and row. With $\alpha = .01$, $df_B = 2$, and $df_W = 12$, $F_{crit} = 6.93$. The decision rule is: *If $F_{obt} > 6.93$, H_0 will be rejected.* The decision rule in ANOVA is always phrased using an inequality with *greater than* because the *F* distribution contains only positive values, so the critical region is always in the right-hand tail.

Step 4: Compute the Obtained Value of the Test Statistic

Step 4 entails a variety of symbols and abbreviations, all of which are listed and defined in Table 12.2. Stop for a moment and study this chart. You will need to know these symbols and what they mean in order to understand the concepts and formulas about to come.

Table 12.2 Elements of ANOVA

Sample Sizes	*Means*	*Sums of Squares*	*Mean Squares*
n_k = the sample size of group *k*; the number of cases in each group	$\bar{x}_k$ = a group mean; each group's mean on the DV	SS_B = between-groups sums of squares	MS_B = between-groups mean squares
N = the total sample size across all groups	$\bar{x}_G$ = the grand mean; the mean for the entire sample regardless of group	SS_W = within-groups sums of squares	MS_W = within-groups mean squares
		SS_T = total sums of squares; $SS_B + SS_W = SS_T$	

You already know that each group has a sample size (n_k) and that the entire sample has a total sample size (N). Each group also has its own mean ($\bar{x}_k$) and the entire sample has a grand mean ($\bar{x}_G$). These sample sizes and means, along with other numbers that will be discussed shortly, are used to calculate the three types of sums of squares. The sums of squares are then used to compute mean squares, which, in turn, are used to derive the obtained value of *F*. We will first take a look at the formulas for the three types of sums of squares: total (SS_T), between-group (SS_B), and within-group (SS_W).

$$SS_T = \sum_i \sum_k x^2 - \frac{\left(\sum_i \sum_k x\right)^2}{N}, \text{ where}$$

> Formula 12(3)

$\sum_i$ = the sum of all scores *i* in group *k*,

$\sum_k$ = the sum of each group total across all groups in the sample,

x = the raw scores, and

N = the total sample size across all groups.

$$SS_B = \sum_k n_k \left(\bar{x}_k - \bar{x}_G \right)^2, \text{ where}$$

Formula 12(4)

n_k = the number of cases in group k,

$\bar{x}_k$ = the mean of group k, and

$\bar{x}_G$ = the grand mean across all groups.

$$SS_w = SS_T - SS_B.$$

Formula 12(5)

The double summation signs in the SS_T formula look a bit intimidating, but they are just instructing you to sum sums. The i subscript denotes individual scores and k signifies groups, so the double sigmas direct you to first sum the scores within each group and to then add up all the group sums to form a single sum representing the entire sample.

The easiest way to compute the sums of squares is to use a table like that below. What we ultimately want from the table are (a) the sums of the raw scores for each group, (b) the sums of each group's *squared* raw scores, and (c) each group's mean. All of these numbers are displayed in Table 12.3.

Table 12.3   ANOVA Computation Table for Asian Property Offender Sentences

| Number of Prior Convictions | | | | | |
No Priors (x_1)	x_1^2	1 – 4 Priors (x_2)	x_2^2	5+ Priors (x_3)	x_3^2
5.00	25.00	11.67	136.19	1.33	1.77
10.00	100.00	5.00	25.00	2.00	4.00
10.00	100.00	5.00	25.00	1.33	1.77
.75	.56	5.00	25.00	5.00	25.00
$n_1 = 4$	$\Sigma x_1^2 = 225.56$	5.00	25.00	5.00	25.00
$\Sigma x_1 = 25.75$		5.00	25.00	$n_3 = 5$	$\Sigma x_3^2 = 57.54$
$\bar{x}_1 = 6.44$		$n_2 = 6$	$\Sigma x_2^2 = 261.19$	$\Sigma x_3 = 14.66$	
		$\Sigma x_2 = 36.67$		$\bar{x}_3 = 2.93$	
		$\bar{x}_2 = 6.11$			

We also need the grand mean, which is computed by summing all of the raw scores across groups and dividing by the total sample size N, as such:

$$\bar{x}_G = \frac{\sum_i \sum_k x}{N}.$$

 Formula 12(6)

Here,

$$\bar{x}_G = \frac{25.75+36.67+14.66}{4+6+5} = \frac{77.08}{15} = 5.14.$$

With all of this information, we are ready to compute the three types of sums of squares, as follows. We start with SS_T:

$$SS_T = (225.56+261.19+57.54) - \frac{(25.75+36.67+14.66)^2}{15} = 544.29 - \frac{77.08^2}{15}$$

$$= 544.29 - \frac{5941.33}{15} = 544.29 - 396.09 = 148.20.$$

Then it is time for the between-groups sums of squares:

$$SS_B = 4(6.44-5.14)^2 + 6(6.11-5.14)^2 + 5(2.93-5.14)^2 = 4(1.30)^2 + 6(.97)^2 + 5(-2.21)^2$$

$$= 4(1.69) + 6(.94) + 5(4.88) = 6.76 + 5.64 + 24.40 = 36.80.$$

Next, we calculate the within-groups sums of squares:

$$SS_W = 148.20 - 36.80 = 111.40.$$

STUDY TIP: NO NEGATIVES IN ANOVA

A great way to help you check your math as you go through Step 4 of ANOVA is to remember that the final answers for any of the sums of squares, mean squares, or F_{obt} will never be negative. If you get a negative number for any of your final answers in Step 4, you will know immediately that you made a calculation error and you should go back and locate the mistake.

We now have what we need to compute the mean squares (symbolized *MS*). Mean squares transform sums of squares into variances by dividing SS_B and SS_w by their respective degrees of freedom, df_B and df_w. The mean squares formulas are

$$MS_B = \frac{SS_B}{k-1} \text{ and} \qquad \text{Formula 12(7)}$$

$$MS_W = \frac{SS_W}{N-k}. \qquad \text{Formula 12(8)}$$

Plugging in our numbers:

$$MS_B = \frac{36.80}{3-1} = 18.40.$$

$$MS_W = \frac{111.40}{15-3} = 9.28.$$

We now have what we need to calculate F_{obt}. The F statistic is the ratio of between-group variance to within-group variance and is computed as

$$F_{obt} = \frac{MS_B}{MS_W}.$$

Formula 12(9)

Inserting the numbers from the present example:

$$F_{obt} = \frac{18.40}{9.28} = 1.98.$$

Step 4 is done! $F_{obt} = 1.98$.

Step 5: Make a Decision About the Null Hypothesis and State the Substantive Conclusion

The decision rule stated that if the obtained value exceeded 6.93, the null would be rejected. With an F_{obt} of 1.98, the null is retained. The substantive conclusion is that there is no significant difference between the groups in terms of sentence length received. In other words, the number of prior convictions possessed by male Asian felony property offenders who are imprisoned upon conviction does not influence the length of the terms of incarceration imposed upon them. The absence of a relationship seems counterintuitive, but remember that this is a bivariate analysis and we have not accounted for important factors such as the severity of those prior convictions. All of these offenders, moreover, pled guilty to the charges against them, and some of these pleas may have been the product of bargains in which prosecutors offered to disregard prior convictions in exchange for the plea of guilty.

We will go through another ANOVA example. If you are not already using your calculator to work through the steps as you read and make sure you can replicate the results obtained here in the book, start doing so. This is an excellent way to learn the material.

Example 2: Are there regional differences in handgun murder rates?

Handguns are a prevalent murder weapon and in some locations, they account for more deaths than all other modalities combined. In criminal justice/criminology researchers' ongoing efforts to learn about violent crime, the question arises as to geographical differences in handgun-involved murders. For the second ANOVA example, we use Uniform Crime Report data to find out whether there are significant regional differences in handgun murder rates (calculated as the number of murders by handgun in 2009 per 100,000 residents in each state). A random sample of states was drawn and the selected states were divided by region. Table 12.4 contains the data in the format that will be used for computations. Alpha will be set at .05.

Table 12.4 Handgun Murder Rate by Region, 2009

Region							
Northeast		Midwest		South		West	
(x_1)	$(x_1)^2$	(x_2)	$(x_2)^2$	(x_3)	$(x_3)^2$	(x_4)	$(x_4)^2$
1.45	2.10	2.12	4.49	4.16	17.31	.14	.02
.30	.09	.10	.01	1.87	3.50	1.09	1.19
.71	.50	1.35	1.82	3.29	10.82	.92	.85
2.17	4.71	.66	.44	2.82	7.95	2.50	6.25
.00	.00	.15	.02	2.52	6.35	1.13	1.28
$\Sigma x_1 = 4.63$	$\Sigma x_1^2 = 7.40$	$\Sigma x_2 = 4.38$	$\Sigma x_2^2 = 6.78$	2.67	7.13	.19	.04
$n_1 = 5$		$n_2 = 5$		1.37	1.88	$\Sigma x_4 = 5.97$	$\Sigma x_4^2 = 9.63$
$\bar{x}_1 = .93$		$\bar{x}_2 = .88$		$\Sigma x_3 = 18.70$	$\Sigma x_3^2 = 54.94$	$n_4 = 6$	
				$n_3 = 7$		$\bar{x}_4 = 1.00$	
				$\bar{x}_3 = 2.67$			

Step 1. State the Null and Alternative Hypotheses

$$H_0: \mu_1 = \mu_2 = \mu_3 = \mu_4$$

$$H_1: some\ \mu_i \neq some\ \mu_j$$

Step 2. Identify the Distribution and Compute the Degrees of Freedom

This being an ANOVA, the F distribution will be employed. There are four groups, so $k = 4$. The total sample size is $N = 5 + 5 + 7 + 6 = 23$. Using Formulas 12(1) and 12(2), the degrees of freedom are:

$$df_B = 4 - 1 = 3$$

$$df_W = 23 - 4 = 19.$$

Step 3. Identify the Critical Value and State the Decision Rule

With $\alpha = .05$ and the above-derived df values, $F_{crit} = 3.13$. The decision rule states that *if $F_{obt} > 3.13$, H_0 will be rejected.*

Step 4: Calculate the Obtained Value of the Test Statistic

We begin by calculating the total sums of squares:

$$SS_T = (7.40 + 6.78 + 54.94 + 9.63) - \frac{(4.63 + 4.38 + 18.70 + 5.97)^2}{23}$$

$$= 78.75 - \frac{33.68^2}{23} = 78.75 - \frac{1134.34}{23} = 78.75 - 49.32 = 29.43.$$

Before computing the between-groups sums of squares, we need the grand mean:

$$\bar{x}_G = \frac{4.63 + 4.38 + 18.70 + 5.97}{23} = \frac{33.68}{23} = 1.46.$$

Now SS_B can be calculated:

$$SS_B = 5(.93 - 1.46)^2 + 5(.88 - 1.46)^2 + 7(2.67 - 1.46)^2 + 6(1.00 - 1.46)^2$$

$$= 5(-.53)^2 + 5(-.58)^2 + 7(1.21)^2 + 6(-.46)^2$$

$$= 5(.28) + 5(.34) + 7(1.46) + 6(.21) = 1.40 + 1.70 + 10.22 + 1.26 = 14.58.$$

Next, we calculate the within-groups sums of squares:

$$SS_w = 29.43 - 14.58 = 14.85.$$

Plugging our numbers into Formulas 12(7) and 12(8) for mean squares:

$$MS_B = \frac{14.58}{4 - 1} = 4.86$$

$$MS_W = \frac{14.85}{23 - 4} = .78.$$

Finally, using Formula 12(9) to derive F_{obt}:

$$F_{obt} = \frac{4.86}{.78} = 6.23.$$

This is the obtained value of the test statistic. $F_{obt} = 6.23$, and Step 4 is complete.

Step 5: Make a Decision About the Null and State the Conclusion

In Step 3, the decision rule stated that if F_{obt} turned out to be greater than 3.13, the null would be rejected. F_{obt} ended up being 6.23, so the null is indeed rejected. The substantive interpretation is that there is a significant difference across regions in the handgun murder rate. Which region or regions differ from which other regions, you ask? The F statistic is silent with respect to the location and number of differences, so post hoc tests are used to get this information. The next section will cover post hoc tests and measures of association that can be used to gauge relationship strength.

▣ WHEN THE NULL IS REJECTED: A MEASURE OF ASSOCIATION AND POST HOC TESTS

If the null is not rejected in ANOVA, then the analysis stops because the conclusion is that the independent and dependent variables are not related. If the null is rejected, however, it is customary to explore the statistically significant results in more detail using *measures of association* (MA) and *post hoc tests*. MAs permit an assessment of the strength of the relationship between the IV and DV, while post hoc tests allow researchers to determine which groups are significantly different from which other ones. The MA that will be discussed here is fairly easy to calculate by hand, but the post hoc tests will be discussed and then demonstrated in the SPSS section, as they are computationally intensive.

Omega squared (ω^2) is an MA for ANOVA that is expressed as the proportion of the total variability in the sample that is due to between-group differences. Omega squared can be left as a proportion or multiplied by 100 to form a percentage. Larger values of ω^2 indicate stronger IV-DV relationships, while smaller values signal weaker associations. Omega squared is computed as

$$\omega^2 = \frac{SS_B - (k-1)MS_w}{MS_W + SS_T}.$$

There are many different types of post hoc tests, so two of the most popular ones are presented here. The first is **Tukey's honest significant difference (HSD)**. Tukey's test compares each group to all the others in a series of two-variable hypothesis tests. The null hypothesis in each comparison is that the two group means are equal; rejection of the null means that there is a significant difference between the means. In this way, Tukey's is conceptually similar to a series of t tests, though the HSD method sidesteps the problem of familywise error.

Omega squared: A measure of association used in ANOVA when the null has been rejected in order to assess the magnitude of the relationship between the independent and dependent variables. This measure shows the proportion of the total variability in the sample that is attributable to between-group differences.

Tukey's honest significant difference: A widely used post hoc test used in ANOVA when the null is rejected as a means of determining the number and location of differences between groups.

Bonferroni: A widely used and relatively conservative post hoc test used in ANOVA when the null is rejected as a means of determining the number and location of differences between groups.

Bonferroni is another commonly used test and owes its popularity primarily to the fact that it is fairly conservative. This means that it minimizes Type I error (erroneously rejecting a true null) at the cost of increasing the likelihood of a Type II error (erroneously retaining a false null). The Bonferroni, though, has been criticized for being too conservative. In the end, the best method is to select both Tukey's and Bonferroni in order to garner a holistic picture of your data and make an informed judgment.

STUDY TIP: WHEN POST HOC TESTS AND MAS ARE APPROPRIATE AND WHEN THEY ARE NOT

Would it be appropriate to compute omega squared and post hoc tests for the ANOVA in Example 1 pertaining to Asian property offenders? Why or why not?

We will now turn to SPSS. You will first learn how to run an ANOVA using this program and the data used in Example 2 above. We will then look at omega squared and post hoc tests.

🖳 SPSS

Let us revisit the question asked in Example 2 regarding whether handgun murder rates vary by region. To run an ANOVA in SPSS, follow the steps depicted in Figure 12.4. Use the *Analyze → Compare Means → One-Way ANOVA* sequence to bring up the dialog box on the left side in Figure 12.2 and then select the variables you want to use. Move the independent variable to the *Factor* space and the dependent variable to the *Dependent List*. Then click *Post Hoc* and select the Bonferroni and Tukey tests. Click *Continue* and *OK* to produce the output shown in Figure 12.5.

The first box of the output shows the results of the hypothesis test. You can see the sums of squares, *df*, and mean squares for within groups and between groups. There are also total sums of squares and total degrees of freedom. The number in the F column is F_{obt}; here, you can see that $F_{obt} = 6.329$. When we did the calculations by hand, we got 6.23. Our hand calculations had some rounding error, but this did not affect the final decision regarding the null because you can also see that the significance value (the p value) is .004, which is less than .05, the value at which α was set. The null hypothesis is rejected in the SPSS context just like it was in the hand calculations.

The next box in the output shows the Tukey and Bonferroni post hoc tests. The difference between these tests is in the p values in the *Sig.* column. In the present case, those differences are immaterial because the results are the same across both types of tests. Based on the asterisks that flag significant results and the fact that the p values associated with the flagged numbers are less than .05, it is apparent that the South is the region that stands out from the others. Its mean is significantly greater than all three of the other regions' means. The Northeast, West, and Midwest do not differ significantly from one another, as evidenced by the fact that all of their p values are greater than .05.

Now it is time to find out just how strong the relationship between region and handgun murder rates is. To find out, we can use ω^2 from Formula 12(10):

$$\omega^2 = \frac{14.67 - (4-1).77}{.77 + 29.36} = \frac{14.67 - 2.31}{30.13} = \frac{12.36}{30.13} = .41.$$

Figure 12.4 Running an ANOVA in SPSS

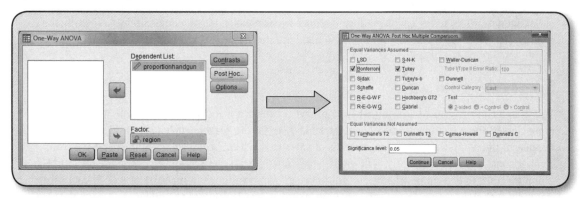

Figure 12.5 ANOVA Output

Handgun murders per 100,000

	Sum of Squares	df	Mean Square	F	Sig.
Between Groups	14.674	3	4.891	6.329	.004
Within Groups	14.684	19	.773		
Total	29.358	22			

Post Hoc Tests

Multiple Comparisons

Dependent Variable:Handgun murders per 100,000

	(I) Region	(J) Region	Mean Difference (I-J)	Std. Error	Sig.	95% Confidence Interval Lower Bound	95% Confidence Interval Upper Bound
Tukey HSD	Northeast	Midwest	.05037	.55600	1.000	-1.5130	1.6138
		South	-1.74371*	.51475	.015	-3.1911	-.2963
		West	-.06898	.53233	.999	-1.5658	1.4278
	Midwest	Northeast	-.05037	.55600	1.000	-1.6138	1.5130
		South	-1.79408*	.51475	.012	-3.2415	-.3467
		West	-.11935	.53233	.996	-1.6162	1.3775
	South	Northeast	1.74371*	.51475	.015	.2963	3.1911
		Midwest	1.79408*	.51475	.012	.3467	3.2415
		West	1.67473*	.48909	.014	.2995	3.0500
	West	Northeast	.06898	.53233	.999	-1.4278	1.5658
		Midwest	.11935	.53233	.996	-1.3775	1.6162
		South	-1.67473*	.48909	.014	-3.0500	-.2995
Bonferroni	Northeast	Midwest	.05037	.55600	1.000	-1.5864	1.6872
		South	-1.74371*	.51475	.019	-3.2591	-.2283
		West	-.06898	.53233	1.000	-1.6361	1.4981
	Midwest	Northeast	-.05037	.55600	1.000	-1.6872	1.5864
		South	-1.79408*	.51475	.015	-3.3095	-.2787
		West	-.11935	.53233	1.000	-1.6865	1.4478
	South	Northeast	1.74371*	.51475	.019	.2283	3.2591
		Midwest	1.79408*	.51475	.015	.2787	3.3095
		West	1.67473*	.48909	.017	.2349	3.1146
	West	Northeast	.06898	.53233	1.000	-1.4981	1.6361
		Midwest	.11935	.53233	1.000	-1.4478	1.6865
		South	-1.67473*	.48909	.017	-3.1146	-.2349

*. The mean difference is significant at the 0.05 level.

Omega squared shows that 41% of the total variability in the states' handgun murder rates is a function of regional characteristics. Region appears to be a very important determinate of the prevalence of handgun murders. In the ongoing effort to reduce firearm violence, it would be useful to identify the characteristics of regions that exert impact on handgun murders.

CHAPTER SUMMARY

This chapter taught you what to do when you have a categorical independent variable with three or more classes and a continuous dependent variable. A series of *t* tests in such a situation is not viable because of the familywise error rate. Analysis of variance (ANOVA) conducts multiple between-group comparisons in a single analysis. ANOVA's *F* statistic compares between-group variance to within-group variance to determine whether or not between-group variance (a measure of true effect) substantially outweighs within-group variance (a measure of error). If it does, the null is rejected; if it does not, the null is retained.

The ANOVA *F*, though, does not indicate the size of the effect, so this chapter introduced you to a measure of association that allows for a determination of the strength of a relationship. This measure is omega squared (ω^2), and it is for use only when the null has been rejected—there is no sense in examining the strength of an IV-DV relationship that you just said does not exist! Omega squared is interpreted as the proportion of the variability in the DV that is attributable to the IV. It can be multiplied by 100 to be interpreted as a percentage rather than a proportion.

The *F* statistic also does not offer information about the location or number of differences between groups. When the null is retained, this is not a problem because a retained null means that there are no differences between groups; however, when the null is rejected, it is desirable to gather more information about which group or groups differ from which others. This is the reason for the existence of post hoc tests. This chapter covered Tukey's HSD and Bonferroni, which are two of the most commonly used post hoc tests in criminal justice/criminology research. Bonferroni is a conservative test, meaning that it is more difficult to reject the null hypothesis of no difference between groups. It is a good idea to run both tests and, if they produce discrepant information, make a reasoned judgment based upon your knowledge of the subject matter and data. Together, measures of association and post hoc tests can help you glean a comprehensive and informative picture of the relationship between the independent and dependent variables.

CHAPTER 12 REVIEW PROBLEMS

1. A researcher wants to know whether judges' gender (measured as *male; female*) affects the severity of sentences they impose upon convicted defendants (measured as *months of incarceration*). Answer the following questions:

   a. What is the independent variable?
   b. What is the level of measurement of the independent variable?
   c. What is the dependent variable?
   d. What is the level of measurement of the dependent variable?
   e. What type of hypothesis test should the researcher use?

2. A researcher wants to know whether judges' gender (measured as *male; female*) affects the types of sentences they impose upon convicted criminal defendants (measured as *jail; prison; probation; fine; other*). Answer the following questions:

   a. What is the independent variable?
   b. What is the level of measurement of the independent variable?
   c. What is the dependent variable?
   d. What is the level of measurement of the dependent variable?
   e. What type of hypothesis test should the researcher use?

3. A researcher wishes to find out whether arrest deters domestic violence offenders from committing future acts of violence against intimate partners. The researcher measures arrest as *arrest; mediation; separation; no action* and recidivism as *number of arrests for domestic violence within the next three years*. Answer the following questions:

   a. What is the independent variable?
   b. What is the level of measurement of the independent variable?
   c. What is the dependent variable?
   d. What is the level of measurement of the dependent variable?
   e. What type of hypothesis test should the researcher use?

4. A researcher wishes to find out whether arrest deters domestic violence offenders from committing future acts of violence against intimate partners. The researcher measures arrest as *arrest; mediation; separation; no action* and recidivism as whether or not these offenders were arrested for domestic violence within the next two years (measured as *arrested; not arrested*). Answer the following questions:

   a. What is the independent variable?
   b. What is the level of measurement of the independent variable?
   c. What is the dependent variable?
   d. What is the level of measurement of the dependent variable?
   e. What type of hypothesis test should the researcher use?

5. A researcher wants to know whether poverty affects crime. The researcher codes neighborhoods as being *lower-class, middle-class,* or *upper-class* and obtains the crime rate for each area (measured as the number of index offenses per 10,000 residents). Answer the following questions:

   a. What is the independent variable?
   b. What is the level of measurement of the independent variable?
   c. What is the dependent variable?
   d. What is the level of measurement of the dependent variable?
   e. What type of hypothesis test should the researcher use?

6. A researcher wants to know whether the prevalence of liquor-selling establishments (such as bars and convenience stores) in neighborhoods affects crime in those areas. The researcher codes neighborhoods as having *0 – 1, 2 – 3, 4 – 5,* or *6+* liquor-selling establishments. The researcher also obtains the crime rate for each area (measured as the number of index offenses per 10,000 residents). Answer the following questions:

   a. What is the independent variable?
   b. What is the level of measurement of the independent variable?

   c. What is the dependent variable?
   d. What is the level of measurement of the dependent variable?
   e. What type of hypothesis test should the researcher use?

7. The Omnibus Crime Control and Safe Streets Act of 1968 requires state and federal courts to report information on all wiretaps sought by and authorized for law enforcement agencies (Duff, 2010). One question of interest to someone studying wiretaps is whether wiretap use varies by crime type; that is, we might want to know whether law enforcement agents use wiretaps with greater frequency in certain types of investigations than in other types. The table below contains data from the U.S. courts website (www.uscourts.gov/Statistics/WiretapReports/WiretapReport2009.aspx) on the number of wiretaps sought by law enforcement agencies in a sample of states. The wiretaps are broken down by offense type, meaning that each number in the table represents the number of wiretap authorizations received by a particular state for a particular offense in 2009. Using an alpha level of .05, test the null hypothesis of no difference between the group means against the alternative hypothesis that at least one group mean is significantly different from at least one other. Use all five steps. If appropriate, compute and interpret omega squared.

Number of Wiretaps Issued per State, by Type, 2009

Homicide and Assault (x_1)	Narcotics (x_2)	Racketeering (x_3)
2	25	1
0	1	1
0	4	0
1	2	3
14	21	0
1	3	0
2	12	0
$n_1 = 7$	$n_2 = 7$	$n_3 = 7$

8. In Chapter 11, we encountered a question about whether multiple homicide offenders (MHOs) "specialize" in killing or whether they, like other types of offenders, tend to commit a wide variety of crimes. Wright, Pratt, and DeLisi (2008) conducted a study on the matter and found that MHOs' criminal histories are quite similar to those of single homicide offenders, indicating an absence of specialization. Let us use the State Court Processing Statistics to examine this issue from a different angle. The table below contains data on the number of prior arrests among three types of felony defendants: Those facing at least two murder charges; those facing one murder charge; and, for the sake of comparison, those facing rape charges. The question is whether the number of prior arrests differs as a function of charge type and, specifically, whether multiple-murder defendants have a significantly different number of prior arrests compared to other types of defendants. Using an alpha level of .01, test the null hypothesis of no difference between means against the alternative hypothesis that at least one defendant group is significantly different from at least one other. Use all five steps. If appropriate, compute and interpret omega squared.

Number of Prior Arrests for Defendants Facing Multiple
Murder, Single Murder, and Rape Charges, 2006

Multiple Murder (x_1)	Single Murder (x_2)	Rape (x_3)
10	1	0
0	10	0
1	3	1
10	1	10
3	5	0
10	3	10
$n_1 = 6$	5	1
	5	3
	10	10
	2	10
	$n_2 = 10$	0
		0
		$n_3 = 12$

9. Many crime and criminal justice system phenomena vary by region in this country (for example, rates of violent crime and of executions). In the ongoing effort to reduce police injuries and fatalities resulting from assaults, one issue is the technology of violence against officers or, in other words, the type of implements offenders use when attacking police. Like other social events, weapon use might vary across regions. The Uniform Crime Reports collect information on weapons used in officer assaults. These data can be used to find out whether the percentage of officer assaults committed with firearms varies by region. The table below contains the data. (Illinois did not contribute data, so this state is excluded.) Using an alpha level of .01, test the null of no difference between means against the alternative that at least one region is significantly different from at least one other. Use all five steps. If appropriate, compute and interpret omega squared.

Percent of Officer Assaults Committed With Firearms, by Region, 2009

Northeast (x_1)	Midwest (x_2)	South (x_3)	West (x_4)
.76	1.05	2.86	3.55
.00	5.28	2.41	4.52
2.65	4.92	3.49	3.64

(Continued)

(Continued)

Northeast (x_1)	Midwest (x_2)	South (x_3)	West (x_4)
.00	.96	2.12	2.29
.23	1.41	3.39	3.88
5.71	1.50	7.79	4.90
1.90	.68	1.69	.68
1.53	4.71	3.18	1.61
8.33	1.36	2.41	8.25
$n_1 = 9$	.00	3.06	3.32
	.70	3.49	.00
	$n_2 = 11$	6.54	4.79
		5.52	3.13
		1.91	$n_4 = 13$
		5.45	
		5.28	
		$n_3 = 16$	

10. An ongoing source of question and controversy in the criminal court system are the possible advantages that wealthier defendants may have over poorer ones, largely as a result of the fact that the former can pay to hire their own attorneys, while the latter must accept the services of court-appointed counsel. There is a common perception that privately retained attorneys are more skilled and dedicated than their publicly appointed counterparts. Let us examine this issue using a sample of rape defendants from the 2006 State Court Processing Statistics data set. The independent variable is *attorney type* and the dependent variable is *days to pretrial release*, which measures the number of days between arrest and pretrial release for those rape defendants who were released pending trial. (Those who did not make bail or were denied bail altogether are not included.) Using an alpha level of .05, test the null of no difference between means against the alternative that at least one region is significantly different from at least one other. Use all five steps. If appropriate, compute and interpret omega squared.

Days Until Release Among a Sample of
Rape Defendants, by Attorney Type, 2006

Public Defender (x_1)	Assigned Counsel (x_2)	Private Attorney (x_3)
2	0	0
42	0	0

(Continued)

(Continued)

Public Defender (x_1)	Assigned Counsel (x_2)	Private Attorney (x_3)
5	6	0
4	51	1
8	5	3
1	5	24
0	$n_2 = 6$	0
$n_1 = 7$		0
		0
		5
		34
		$n_3 = 11$

11. Does the gender of inmates housed in a prison affect the inmate-on-inmate assault rate in that facility? The table below contains a random sample of prisons drawn from the Census of State and Federal Correctional Facilities. Using an alpha level of .05, test the null of no difference between means against the alternative that at least one facility type is significantly different from at least one other. Use all five steps. If appropriate, compute and interpret omega squared.

Assault Rates by Facility Type

Male Only (x_1)	Female Only (x_2)	Both (x_3)
.00	.00	.09
6.57	.00	.17
.00	1.91	.00
.17	10.64	.00
.21	.00	.00
1.16	2.10	1.67
1.07	3.19	41.53
2.96	.00	.00
.00	7.27	.28
$n_1 = 9$	9.51	.00
	$n_2 = 10$	3.57
		$n_3 = 11$

12. Does the number of juveniles held in an adult correctional facility vary according to the gender of inmates housed in that prison? The table below contains a random sample of prisons drawn from the Census of State and Federal Correctional Facilities. Using an alpha level of .01, test the null of no difference between means against the alternative that at least one facility type is significantly different from at least one other. Use all five steps. If appropriate, compute and interpret omega squared.

Juveniles by Facility Type

Male Only (x_1)	Female Only (x_2)	Both (x_3)
1.00	15.00	1.00
.00	.00	1.00
15.00	.00	.00
.00	.00	.00
7.00	1.00	8.00
1.00	.00	2.00
3.00	4.00	.00
2.00	.00	.00
$n_1 = 8$	1.00	$n_3 = 8$
	3.00	
	$n_2 = 10$	

13. Let us expand upon the question asked in Question 10 above by examining defendants facing drug charges rather than rape charges. The data set *SCPS for Chapter 12.sav* (http://www.sagepub.com/gau) contains State Court Processing Statistics data on all drug defendants in the 2006 sample. The variables are *attorney* and *days* and measure, respectively, the type of attorney a defendant had and the number of days that defendant was held before obtaining pretrial release. The sample has been narrowed to those drug defendants who did obtain pretrial release.

   a. Using SPSS, run an ANOVA with *attorney* as the IV and *days* as the DV. Select the appropriate post hoc tests.
   b. Identify the obtained value of F.
   c. Would you reject the null at an alpha of .01? Why or why not?
   d. State your substantive conclusion about whether there is a relationship between attorney type and days to release for drug defendants.
   e. If appropriate, interpret the post hoc tests to identify the location and total number of significant differences.
   f. If appropriate, compute and interpret omega squared.

14. In Questions 10 and 13, you determined whether there was a relationship between attorney type and days to pretrial release among a sample of adult criminal defendants. Now, using SPSS, do the same for juvenile defendants. The data set *JDCC for Chapter 12.sav* (**http://www.sagepub.com/gau**) contains juvenile defendants facing drug charges. The variables *attorney* and *days* are the same as those used in Question 10.

    a. Using SPSS, run an ANOVA with *attorney* as the IV and *days* as the DV. Select the appropriate post hoc tests.

    b. Identify the obtained value of *F*.

    c. Would you reject the null at an alpha of .01? Why or why not?

    d. State your substantive conclusion about whether there is a relationship between attorney type and days to release for juvenile defendants.

    e. If appropriate, interpret the post hoc tests to identify the location and total number of significant differences.

    f. If appropriate, compute and interpret omega squared.

KEY TERMS

Analysis of Variance (ANOVA)	Within-group variance	Omega squared
Familywise error	*F* statistic	Tukey's honest significant difference
Groups	*F* distribution	
Between-group variance	Post hoc tests	Bonferroni

GLOSSARY OF SYMBOLS AND ABBREVIATIONS INTRODUCED IN THIS CHAPTER

F	The statistic and sampling distribution for ANOVA
n_k	The sample size of group k
N	The total sample size across all groups
$\bar{x}_k$	The mean of group k
$\bar{x}_G$	The grand mean across all cases in all groups
SS_B	Between-groups sums of squares; a measure of true group effect
SS_W	Within-groups sums of squares; a measure of error
SS_T	Total sums of squares; equal to $SS_B + SS_w$

(Continued)

(Continued)

MS_B	Between-group mean square; the variance between groups and a measure of true group effect
MS_W	Within-group mean square; the variance within groups and a measure of error
ω^2	A measure of association that indicates the proportion of the total variability that is due to between-group differences

REFERENCES

Bureau of Justice Statistics. (2006). *State court processing statistics, 1990–2006: Felony defendants in large urban counties.* Ann Arbor, MI: Inter-University Consortium for Political and Social Research.

Duff, J. C. (2010). Report of the director of the Administrative Office of the United States Courts. Retrieved from www.uscourts.gov/uscourts/Statistics/WiretapReports/2009/2009Wiretaptext.pdf

Franklin, T. W., & Fearn, N. E. (2010). Sentencing Asian offenders in state courts: The influence of a prevalent stereotype. *Crime & Delinquency.* Prepublished November 7, 2010. DOI: 10.1177/0011128710386200

Wright, K. A., Pratt, T. C., & DeLisi, M. (2008). Examining offending specialization in a sample of male multiple homicide offenders. *Homicide Studies*, 12(4), 381–398.

Chapter 13

Hypothesis Testing With Two Continuous Variables

Correlation

Thus far, we have learned the hypothesis tests for use when the two variables under examination are both categorical (chi-square), when the independent variable (IV) is categorical and the dependent variable (DV) is a proportion (two-population z test for proportions), when the IV is a two-class categorical measure and the DV is continuous (t tests), and when the IV is categorical with three or more classes and the DV is continuous (ANOVA).

In the current chapter, we will address the technique that is proper when both of the variables are continuous. This technique is **Pearson's correlation**, because it was developed by Karl Pearson, who was instrumental in advancing the field of statistics.

Pearson's correlation: The bivariate statistical analysis used when the independent and dependent variables are both continuous.

Positive correlation: When a one-unit increase in the independent variable produces an increase in the dependent variable.

Negative correlation: When a one-unit increase in the independent variable produces a reduction in the dependent variable.

The question asked in a correlation analysis is, "When the IV increases by one unit, what happens to the DV?" The DV might also increase (a **positive correlation**), it might decrease (a **negative correlation**), or it might do nothing at all (no relationship). Figure 13.1 depicts these possibilities.

Figure 13.1 Three Types of Correlations Between a Continuous IV and a Continuous DV

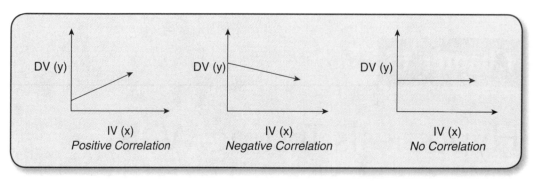

The bivariate associations represented by correlations are **linear relationships**. This means that the amount of change in the DV that is produced by a one-unit increase in the IV remains constant across all levels of the IV. If the correlation between two variables is .50, for instance, then a one-unit increase in the IV is associated with a .50-unit increase in the DV irrespective of whether the IV increases from, for example, 0 to 1 or from 59 to 60. Linear relationships can be contrasted to non-linear or curvilinear relationships such as those displayed in Figure 13.2. You can see in this figure how a one-unit change in the IV produces varying changes in the DV. Sometimes the DV increases, sometimes it decreases, and sometimes it does nothing at all. These nonlinear relationships cannot be modeled using correlational analyses.

Linear relationship: A relationship wherein the change in the dependent variable that is produced by a one-unit increase in the independent variable remains static or constant at all levels of the independent variable.

r **coefficient:** The test statistic in a correlation analysis.

The statistic representing correlations is called the *r* **coefficient**. This coefficient ranges from −1.00 to +1.00. The population correlation coefficient is ρ, which is the Greek letter *rho* (pronounced "roe").

Figure 13.2 Examples of Nonlinear Relationships

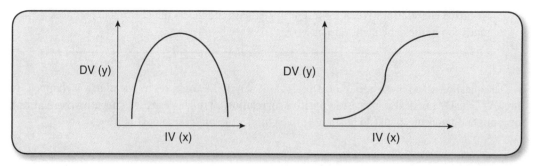

A correlation of ± 1.00 signals a perfect relationship where a one-unit increase in x (the IV) produces an exact one-unit change in y (the DV). Correlations of zero indicate that there is no relationship between the two variables. Coefficients less than zero signify negative relationships, while coefficients greater than zero represent positive relationships. For an example of the use of correlation in criminal justice/ criminology research, see Research Example 13.1.

RESEARCH EXAMPLE 13.1

Part 1: Is perceived risk of Internet fraud victimization related to online purchases?

Many researchers have addressed the issue of perceived risk with regard to people's behavioral adaptations. Perceived risk has important consequences at both the individual and the community levels, because people who believe their likelihood of victimization to be high are less likely to connect with their neighbors, less likely to utilize public spaces in their communities, and more likely to stay indoors to avoid the frightening environment of the outside world. What has not been addressed with much vigor in the criminal justice/criminology literature is the issue of perceived risk of Internet theft victimization. Given how integral the Internet is to American life and the enormous volume of commerce that takes place online every year, it is important to study the online shopping environment as an arena ripe for theft and fraud.

Reisig, Pratt, and Holtfreter (2009) examined this issue in the context of Internet theft victimization. They used self-report data from a survey administered to a random sample of citizens. Their research question was whether perceived risk of Internet theft victimization would dampen people's tendency to shop online because of the vulnerability created when credit cards are used to make Internet purchases. They also examined whether people's financial impulsivity (the tendency to spend money rather than save it and to possibly spend more than one's income provides for) affected perceived risk. Reisig et al. ran a correlation analysis. Was their hypothesis supported? We will revisit this study later in the chapter to find out.

Correlation analyses employ the t distribution because this probability distribution adequately mirrors the sampling distribution of r at small and large sample sizes (see Figure 13.3). The method for conducting a correlation analysis is to first calculate r and then test for the statistical significance of r by comparing t_{crit} and t_{obt}. Keep this two-step procedure in mind so that you understand the analytic technique in Step 4.

Example 1: Is prior criminal record related to in-prison offending?

The behavior of prisoners while they are incarcerated is of concern because of the danger that misbehavior can pose to the safety of staff and other inmates. One question that might be asked is whether

Figure 13.3   The Sampling Distribution of
Correlation Coefficients

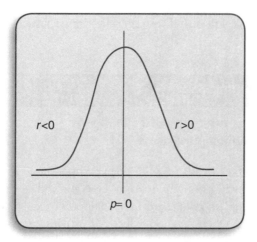

inmates' criminal histories are related to their in-prison behavior. Let us test for a correlation between criminal history and in-prison behavior using the Bureau of Justice Statistics' 2004 *Survey of Inmates in Federal Correctional Facilities* (SIFCF; see Data Sources 13.1). Criminal history is measured as the total number of times each person has been incarcerated during his life, and prison misbehavior is measured as the number of disciplinary write-ups each inmate has received during his current term of incarceration. Both of these variables are continuous. The sample consists of Latino males born in the United States who were serving time for violent offenses. Table 13.1 contains the data. We will use an alpha level of .01 and test the null hypothesis of no correlation against the alternative hypothesis that the two variables are correlated.

Table 13.1   Number of Lifetime Incarcerations and Number of
Disciplinary Infractions Among a Sample of Latino Males

Person	Incarcerations (x)	Disciplinary Reports (y)
A	14	4
B	1	3
C	2	8
D	5	1
E	2	3
N = 5		

DATA SOURCES 13.1

THE SURVEY OF INMATES IN FEDERAL CORRECTIONAL FACILITIES (SIFCF)

The SIFCF (Bureau of Justice Statistics, 2004) contains data gathered from face-to-face interviews with a nationally representative sample of inmates held in federal correctional facilities. Participants are selected in a two-stage sampling design where a random sample of federal prisons is drawn first, and then a random sample of inmates is chosen from within each of the selected facilities. Inmates are asked questions about the offense that led to their current incarceration, prior offenses and prior incarcerations, any abuse or other victimization they may have suffered as children, their history of drug and alcohol use, and their plans for reintegrating into "free" life after their current sentence expires and they are released.

Step 1: State the Null and Alternative Hypotheses

In a correlation analysis, the null hypothesis is that there is no correlation between the two variables. The null is phrased in terms of ρ, the population correlation coefficient. Recall that a correlation coefficient of zero signifies an absence of a relationship between two variables; therefore, the null in correlation is

$$H_0: \rho = 0.$$

There are three options available for the phrasing of the alternative hypothesis. Since correlations use the *t* distribution, these three options are the same as those in *t* tests. There is a two-tailed option (phrased as $H_1: \rho \neq 0$) that predicts a correlation of unspecified direction. This is the option used when a researcher does not wish to make an a priori prediction about whether the correlation is positive or negative. There are also two one-tailed options. The first predicts that the correlation is negative ($H_1: \rho < 0$) and the second predicts that it is positive ($H_1: \rho > 0$).

In the present example, no direction was specified. The alternative hypothesis is merely that the variables are correlated—no attempt was made to predict whether that correlation is positive or whether it is negative. This is, therefore, a two-tailed test and the alternative is

$$H_0: \rho \neq 0.$$

Step 2: Identify the Distribution and Compute the Degrees of Freedom

The *t* distribution is the probability curve used in correlation analyses. This curve is symmetric and, unlike the χ^2 and *F* distributions, has both a positive and a negative side. The degrees of freedom in correlation are computed as

$$df = N - 2.$$

Formula 13(1)

In the present example, there are five people in the sample and the *df* are thus

$$df = 5 - 2 = 3.$$

Step 3: Identify the Critical Value and State the Decision Rule

With a two-tailed test with $\alpha = .01$ and $df = 3$, the absolute value of t_{crit} is 5.841. As this is a two-tailed test, you should remember from the *t* test chapter that there are actually two critical values, one negative and one positive. The critical value of *t* is therefore ±5.841. The decision rule is: *If t_{obt} is either > 5.841 or < −5.841, H_0 will be rejected.*

Step 4: Compute the Obtained Value of the Test Statistic

There are two parts to Step 4 in correlation analyses: first, the correlation coefficient *r* is calculated; and second, the statistical significance of *r* is tested by plugging *r* into the t_{obt} formula. The formula for *r* looks complex, but we will solve it step by step.

$$r = \frac{N \sum xy - \sum x \sum y}{\sqrt{\left[N \sum x^2 - \left(\sum x \right)^2 \right]\left[N \sum y^2 - \left(\sum y \right)^2 \right]}}$$

Formula 13(2)

The formula requires several different sums: $\sum x$, $\sum y$, $\sum xy$, $\sum x^2$, and $\sum y^2$. The easiest way to obtain these sums is to use a table. Table 13.2 reproduces the raw data from Table 13.1 and adds three columns to the right that allow us to compute the needed sums.

Table 13.2   Correlation Computation Table for Example 1

Person	Incarcerations (x)	Disciplinary Reports (y)	xy	x^2	y^2
A	14	4	56	196	16
B	1	3	3	1	9
C	2	8	16	4	64
D	5	1	5	25	1
E	2	3	6	2	9
$N = 5$	$\sum x = 24$	$\sum y = 19$	$\sum xy = 86$	$\sum x^2 = 228$	$\sum y^2 = 99$

Enter the sums into Formula 13(2):

$$r = \frac{5(86) - (24)(19)}{\sqrt{\left[5(228) - 24^2 \right]\left[5(99) - 19^2 \right]}} = \frac{430 - 456}{\sqrt{[1140 - 576][495 - 361]}} = \frac{-26}{\sqrt{(564)(134)}} = \frac{-26}{\sqrt{75576}} = \frac{-26}{274.91} = -.09.$$

The first part of Step 4 is thus complete. We now know that $r = -.09$; however, we do not yet know whether the null hypothesis will be rejected because we have not computed t_{obt}. To make a decision about the null, the following equation is used:

$$t_{obt} = r\sqrt{\frac{N-2}{1-r^2}}.$$

Formula 13(3)

Plugging our numbers in:

$$t_{obt} = -.09\sqrt{\frac{5-2}{1-(-.09)^2}} = -.09\sqrt{\frac{3}{1-.01}} = -.09\sqrt{\frac{3}{.99}} = -.09\sqrt{3.03} = -.09(1.74) = -.16.$$

Step 4 is complete! We know that $r = -.09$ and $t_{obt} = -.16$.

Step 5: Make a Decision About the Null and State the Substantive Conclusion

In the decision rule, we said that the null would be rejected if t_{obt} were either greater than 5.841 or less than −5.841; however, t_{obt} clearly meets neither of these conditions, so the null is retained. The conclusion is that among Latino males incarcerated for a violent offense, there is no correlation between the number of times these inmates have been incarcerated during their lives and the amount of trouble they get into while in prison. Another way of saying this is that as the number of incarcerations a person has experienced increases by one unit, the number of disciplinary reports filed does *not* increase or decrease in a clear, predictable fashion. Knowing about someone's prior criminal history with regard to the number of times he has been incarcerated does not offer information about his conduct during his present term of imprisonment.

STUDY TIP: DO THE CALCULATIONS AS YOU READ

Are you following along with your calculator? If not, start! Doing the calculations yourself to confirm that you arrive at the same answer presented in the text is an invaluable learning tool.

Example 2: Is early-onset offending a predictor of high-rate offending?

At the core of many criminological theories about adult criminal offending is the assumption that offenders who get involved in crime or delinquency at a young age are more likely to display a pattern of antisocial behavior throughout their lives relative to those persons who are older when they commit their first offense. The age at which a person starts engaging in delinquent or criminal acts is called the age of onset. There should be a negative correlation between offenders' age of onset and the total number of offenses they commit—as age of onset increases, the number of offenses should decline.

We will test this hypothesis using data from the SIFCF (Data Sources 13.1). The sample consists of female inmates incarcerated for violent offenses and was narrowed to those persons who were at least 31 years of age at the time the data were collected. This restriction ensures that respondents' criminal histories have had time to manifest. Age of onset is measured as the age at which inmates were arrested for the first time, and lifetime offending is measured as the total number of times they had been arrested. Table 13.3 contains the data in the format that will be used for calculations. Using $\alpha = .05$, we will test the null hypothesis of no correlation against the alternative that the variables are negatively correlated.

Step 1: State the Null and Alternative Hypotheses

$$H_0: \rho = 0$$

$$H_1: \rho < 0$$

Table 13.3 Age at First Arrest and Total Number of Arrests Over Life in a Sample of Female Federal Inmates Incarcerated for Violent Offenses

Person	Age at First Arrest (x)	Lifetime Arrests (y)	xy	x^2	y^2
A	28	4	112	784	16
B	33	3	99	1089	9
C	31	3	93	961	9
D	43	2	86	1849	4
E	15	11	165	225	121
F	47	3	141	2209	9
G	21	21	441	441	441
H	40	3	120	1600	9
I	18	1	18	324	1
J	10	6	60	100	36
K	66	2	132	4356	4
L	18	4	72	324	16
M	15	2	30	225	4
N	26	2	52	676	4
O	15	11	165	225	121
P	27	6	162	729	36
Q	29	2	58	841	4
R	12	21	252	144	441
S	12	21	252	144	441
T	19	4	76	361	16
U	42	2	84	1764	4
V	13	61	793	169	3721
W	28	2	56	784	4
X	26	3	78	676	9
Y	27	3	81	729	9
N = 25	$\Sigma x = 661$	$\Sigma y = 203$	$\Sigma xy = 3678$	$\Sigma x^2 = 21729$	$\Sigma y^2 = 5489$

Step 2: Identify the Distribution and Compute the Degrees of Freedom

The distribution is t, and the df are computed using Formula 13(1):

$$df = 25 - 2 = 23.$$

Step 3. Identify the Critical Value and State the Decision Rule

With a one-tailed test, an alpha of .05, and $df = 23$, $t_{crit} = -1.714$. The critical value is negative because the alternative hypothesis predicts that the correlation is less than zero. The decision rule is: *If* $t_{obt} < -1.714$, H_0 *will be rejected.*

Step 4: Compute the Obtained Value of the Test Statistic

Using Formula 13(2) and the sums from Table 13.3,

$$r = \frac{25(3678) - (661)(203)}{\sqrt{\left[25(21729) - 661^2\right]\left[25(5489) - 203^2\right]}} = \frac{91950 - 134183}{\sqrt{[543225 - 436921][137225 - 41209]}} = \frac{-42233}{\sqrt{(106304)(96016)}}$$

$$= \frac{-42233}{\sqrt{10206884864}} = \frac{-42233}{101029.13} = -.42.$$

Now r is entered into Formula 13(3),

$$t_{obt} = -.42\sqrt{\frac{25-2}{1-(-.42)^2}} = -.42\sqrt{\frac{23}{1-.18}} = -.42\sqrt{\frac{23}{.82}} = -.42\sqrt{28.05} = -.42(5.30) = -2.23.$$

And Step 4 is done! $t_{obt} = -2.23$.

Step 5: Make a Decision About the Null and State the Substantive Conclusion

The decision rule stated that the null would be rejected if t_{obt} was less than -1.714, and since -2.23 meets this criterion, the null is indeed rejected—r is statistically significant. The substantive conclusion is that among female offenders incarcerated for violent offenses, there is a negative correlation between age of onset and lifetime arrests. As the age of first arrest increases by one year, the number of lifetime arrests declines.

The absolute value of r represents the proportion of a one-unit change in the DV that accompanies a one-unit increase in the IV. In Example 2, the r value of $-.42$ means that a one-unit increase in age of onset corresponds to a reduction in lifetime arrests of $.42$, or just under one half of one arrest. As with all other bivariate relationships, we have to limit our conclusion about these two variables to saying merely that they are related; there are many factors other than age of onset that impact lifetime offending, so it would be faulty to leap to the conclusion that age of onset alone exerts substantial causal impact on lifetime offending.

RESEARCH EXAMPLE 13.2

Do good recruits make good cops?

The purpose of police training academies is to prepare recruits for work on the street; however, it is not known to what extent academy training relates to on-the-job performance. Henson, Reyns, Klahm, and Frank (2010) attempted to determine whether and to what extent recruits' academy performance is associated with their later effectiveness as police officers. They used three dependent variables: the evaluations new

(Continued)

(Continued)

officers received from their supervisors; the number of complaints that were lodged against those new officers; and the number of commendations the officers received for exemplary actions. The independent variables consisted of various measurements of academy performance. Henson et al. obtained the following correlations. Note that the authors did not report correlation coefficients that were not statistically significant; these coefficients are labeled *ns* for *nonsignificant*.

	Evaluation	Complaints	Commendations
Civil Service Exam	ns	−.14**	ns
Physical Agility Rating	.19*	.13*	ns
Overall Academy Score	.12*	ns	.12*

Source: Adapted From Table 3 in Henson, Reyns, Klahm, and Frank (2010).

*p < .01;**p < .001.

Were the authors' predictions supported? Is academy performance related to the quality of the job those recruits do once they are on the street? The results were mixed. You can see in the table that many of the correlations were statistically significant, but nearly as many were not. It would appear from this analysis that academy performance and on-the-job performance are not as closely related as would be expected or hoped. These findings have implications for police training procedures.

◫ BEYOND STATISTICAL SIGNIFICANCE: SIGN, MAGNITUDE, AND COEFFICIENT OF DETERMINATION

When the null hypothesis is rejected in a correlation hypothesis test, the correlation coefficient r can be examined with respect to its substantive meaning. A rejected null is not in and of itself an indication that the variables are strongly correlated. The null can be rejected even when a correlation is of little practical importance. When the null is rejected, criminal justice/criminology researchers turn to three interpretive measures to assess the substantive importance of a statistically significant r: *sign*, *magnitude*, and *coefficient of determination*.

The sign of the correlation coefficient indicates whether the correlation between the IV and DV is negative or positive. Take another look at Figure 13.1 to refresh your memory as to what negative and positive correlations look like. The magnitude is an evaluation of the strength of the relationship based on the value of r. There are no set-in-stone rules for determining whether a given r value is strong, moderate, or weak in magnitude; this judgment is based upon a researcher's knowledge of the subject matter. The diagram in Figure 13.4 offers general guidelines that can be used as benchmarks; however, rules about what is considered strong, moderate, or weak may vary across different areas of study.

Figure 13.4   General Guidelines for Interpreting the Magnitude of Correlation Coefficients

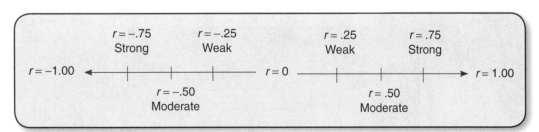

The coefficient of determination is calculated as r^2. The result is a proportion that can be converted to a percentage and interpreted as the percentage of the variance in the DV that is attributable to the IV. As a percentage, the coefficient ranges from zero to 100, with higher numbers signifying stronger relationships and numbers closer to zero representing weaker associations.

Let us interpret the sign, magnitude, and coefficient of determination for the correlation coefficient computed in Example 2. Since we retained the null in Example 1, we cannot apply the three interpretive measures to the r value obtained in this problem. The null was rejected in Example 2, though, so the correlation coefficient can be interpreted. Recall that in this problem, $r = -.42$.

The sign of the correlation coefficient is negative, meaning that a one-unit increase in the IV is associated with a reduction in the DV (a reduction of .42 of one unit, specifically). The magnitude is close to, though a little shy of, the $-.50$ benchmark depicted in Figure 13.2. We can conclude that age of onset is moderately related to lifetime arrests. Finally, the coefficient of determination is $r^2 = (-.42)^2 = .18$, which means that 18% of the variance in lifetime arrests is attributable to the age of onset. This underscores the moderate nature of the correlation—18% is certainly nothing to scoff at and does indicate that age of onset is an important predictor of lifetime arrests, but it also leaves 82% of the variance in arrests unaccounted for. There is a clear need to discover more independent variables, because age of onset is an important but not the lone force behind adult criminality.

▣ SPSS

Correlations are run in SPSS using the *Analyze* → *Correlation* → *Bivariate* sequence. Once the dialog box shown in Figure 13.5 appears, select the variables of interest and move them into the analysis box as shown. Then click *OK*. The variables shown in Figure 13.5 are those that were used in Example 1. Figure 13.6 shows the output.

The result is what is called a *correlation matrix*, meaning that it is split by what is called a *diagonal* (here, the cells in the upper left and lower right corners of the matrix) and is symmetrical on both of the off-diagonal sides. The numbers in the diagonal are always a perfect 1.00 because they represent each variable's correlation with itself. The numbers in the off-diagonals are the ones to look at. The number associated with Pearson's correlation is the value of r. In Figure 13.4, you can see that $r = -.094$, which if rounded to two decimal places is identical to the value of r we reached by hand in Example 1.

Figure 13.5 Running a Correlation in SPSS

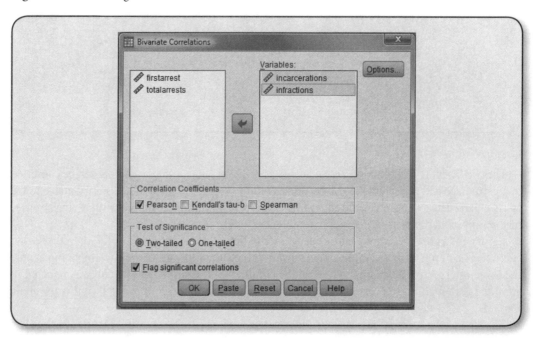

Figure 13.6 SPSS Correlation Output

Correlations

		Example 1: Total lifetime incarcerations	Example 1: Number of disciplinary reports received
Example 1: Total lifetime incarcerations	Pearson Correlation	1	-.094
	Sig. (2-tailed)		.881
	N	5	5
Example 1: Number of disciplinary reports received	Pearson Correlation	-.094	1
	Sig. (2-tailed)	.881	
	N	5	5

The *Sig.* value is, as always, the obtained significance level or *p* value. This number is compared to alpha to determine whether or not the null will be rejected: If $p < \alpha$, the null is rejected; if $p > \alpha$, the null is retained. In this example, $\alpha = .01$ and $p = .881$, so the null is retained. The conclusion is that the two variables are not correlated. Again, this is the same result we arrived at by hand.

RESEARCH EXAMPLE 13.1

Part 2

Recall that Reisig et al. (2009) predicted that people's perceived likelihood of falling victim to Internet theft would lead to less frequent engagement in the risky practice of purchasing items online using credit cards. They also thought that financial impulsivity, as an indicator of low self-control, would affect people's perceptions of risk. Below is an adaptation of the correlation matrix obtained by these researchers.

	Perceived Risk	*Financial Impulsivity*	*Online Purchases*
Perceived Risk	1.00		
Financial Impulsivity	.11*	1.00	
Online Purchases	−.12*	.05	1.00

Source: Adapted From Table 1 in Reisig et al. (2009).
*$p < .05$.

Were the researchers' hypotheses correct? The results were mixed. On one hand, they were correct in that the correlations between the IVs and the DV were statistically significant. You can see the significance of these relationships indicated by the asterisks that flag both of these correlations as being statistically significant at an alpha level of .05. Since the null was rejected, it is appropriate to interpret the coefficients. Regarding sign, perceived risk was negatively related to online purchases (that is, greater perceived risk meant less online purchasing activity) and financial impulsivity was positively related to perceived risk (in other words, financially impulsive people were likely to see themselves as facing an elevated risk of victimization).

The reason the results were mixed, though, is that the correlations—though statistically significant—were not strong. Using the magnitude guidelines in Figure 13.2, it can be seen that −.12 and .11 are very weak. The coefficient of determination for each one is $(-.12)^2 = .01$ and $.11^2 = .01$, so only 1% of the variance in online purchases was attributable to perceived risk, and only 1% was due to financial impulsivity, respectively. This illustrates the potential discrepancy between statistical significance and substantive importance—both of these correlations were statistically significant, but neither meant much in terms of substantive or practical implications. As always, though, it must be remembered that these analyses were bivariate and that the addition of more IVs may alter the IV-DV relationships observed here.

CHAPTER SUMMARY

This chapter introduced you to Pearson's correlation, the hypothesis-testing technique that is appropriate in studies employing two continuous variables. The correlation coefficient is symbolized r and ranges from -1.00 to $+1.00$. The value of r represents the amount of linear change in the DV that accompanies a one-unit increase in the IV. Values of r approaching ± 1.00 are indicative of strong relationships, while values close to zero signify weak correlations. The statistical significance of r is tested using the t distribution because the sampling distribution of r is normal in shape.

When the null hypothesis is rejected in a correlation test, the sign, magnitude, and coefficient of determination should be examined in order to determine the substantive, practical meaning of r. The sign indicates whether the correlation is positive or negative. The magnitude can be assessed using the general guidelines laid out in Figure 13.3. The coefficient of determination is computed by squaring r to produce a proportion (or percentage, when multiplied by 100) that represents the amount of variance in the dependent variable that is attributable to the independent variable.

Correlation analyses can be run in SPSS. The program provides the calculated value of the correlation coefficient r and its associated significance value (its p value). When p is less than alpha, the null is rejected and the correlation coefficient can be interpreted for substantive meaning; when p is greater than alpha, the null is retained and r is not interpreted because the conclusion is that the two variables are not correlated.

CHAPTER 13 REVIEW PROBLEMS

1. A researcher wishes to test the hypothesis that parental incarceration is a risk factor for lifetime incarceration of the adult children of incarcerated parents. She measures parental incarceration as *had a parent in prison; did not have a parent in prison* and adult children's incarceration as *incarcerated; not incarcerated*.

   a. Identify the independent variable.
   b. Identify the level of measurement of the independent variable.
   c. Identify the dependent variable.
   d. Identify the level of measurement of the dependent variable.
   e. What type of inferential analysis should the researcher run to test for a relationship between these two variables?

2. A researcher wishes to test the hypothesis that coercive control can actually increase, rather than reduce, subsequent criminal offending among persons who have been imprisoned. He measures coercive control as *number of years spent in prison* and recidivism as *number of times rearrested after release from prison*.

   a. Identify the independent variable.
   b. Identify the level of measurement of the independent variable.
   c. Identify the dependent variable.
   d. Identify the level of measurement of the dependent variable.
   e. What type of inferential analysis should the researcher run to test for a relationship between these two variables?

3. A researcher thinks that there may be a relationship between neighborhood levels of socioeconomic disadvantage and the amount of violent crime that occurs in those neighborhoods. She measures socioeconomic disadvantage as the percentage of neighborhood residents that live below the poverty

line, and she measures violent crime as the number of violent offenses reported to police per 1,000 neighborhood residents.

a. Identify the independent variable.
b. Identify the level of measurement of the independent variable.
c. Identify the dependent variable.
d. Identify the level of measurement of the dependent variable.
e. What type of inferential analysis should the researcher run to test for a relationship between these two variables?

4. A researcher hypothesizes that a new policing strategy involving community meetings designed to educate residents on the importance of self-protective measures will increase the use of such measures. He gathers a sample of local residents and finds out whether they have participated in an educational session (*have participated; have not participated*). Among each group, he computes the proportion of residents who now take self-protective measures.

a. Identify the independent variable.
b. Identify the level of measurement of the independent variable.
c. Identify the dependent variable.
d. Identify the level of measurement of the dependent variable.
e. What type of inferential analysis should the researcher run to test for a relationship between these two variables?

5. Explain what it means for a relationship to be linear.

6. Explain what it means for a relationship to be nonlinear.

7. Is there a correlation between the amount of money states spend on prisons and those states' violent crime rates? There could be a negative correlation insofar as prisons may suppress crime, in which case money invested in incarceration produces a reduction in violence; however, there could also be a positive correlation if prison expenditures do not reduce crime but, rather, merely reflect the amount of violence present in a state. Morgan, Morgan, and Boba (2010) offer information on the dollars spent per capita on prison expenditures in 2009, and the UCR provides 2009 violent crime data. The table below contains dollars spent per capita and violent crime rates per 1,000 citizens for a random sample of five states. Using an alpha level of .05, test the null hypothesis of no correlation against the alternative that the variables are correlated. Use all five steps. If appropriate, interpret the sign, magnitude, and coefficient of determination.

State	Prison Dollars Per Capita (x)	Violent Crime Rate (y)
Florida	140	6.12
Idaho	135	2.28
Maine	99	1.20
Nebraska	120	2.82
Texas	133	4.91
N = 5		

8. One aspect of deterrence theory predicts that punishment is most effective at preventing crime when would-be criminals know that they stand a high likelihood of being caught and penalized for lawbreaking. When certainty breaks down, by contrast, crime rates may increase because offenders feel more confident that they will not be apprehended. The table below contains Uniform Crime Report (UCR) data on the clearance rate for property crimes in 2008 and the property crime rate per 100 in 2009. The prediction is that higher clearance rates in one year will be associated with lower property crime rates the following year. Using an alpha of .01, test the null hypothesis of no correlation against the alternative that the two variables are negatively correlated. Use all five steps. If appropriate, interpret the sign, magnitude, and coefficient of determination.

Region	Percent of Property Crimes Cleared (x)	Property Crime Rate (y)
New England	15.40	2.33
Mid-Atlantic	22.80	2.05
East North Central	16.30	2.93
West North Central	19.40	2.84
South Atlantic	19.10	3.50
East South Central	17.60	3.34
West South Central	15.80	3.92
Mountain	16.90	3.10
Pacific	14.50	2.90
$N = 9$		

9. Deterrence theory would predict that greater police presence in a particular area would cause would-be offenders to perceive a high likelihood of apprehension and therefore to not commit crime. This would suggest that the relationship between the number of police agencies in an area and the number of crimes in that same area would be negative. On the other hand, it is a criminological truism that crime is concentrated in specific geographic areas. Police agencies would also then be concentrated in certain areas because high-crime locations require greater investments in public safety. From this perspective, the correlation between crime and police agencies should be positive. We will analyze this issue using data from CQ Press (Morgan, Morgan, & Boba, 2010). The independent variable is *crimes per square mile* and the dependent variable is *law enforcement agencies per 1,000 square miles.* Using an alpha level of .05, test the null hypothesis that there is no correlation between these variables against the alternative hypothesis that they are positively correlated. Use all five steps. If appropriate, interpret the sign, magnitude, and coefficient of determination.

State	Crime per Sq. Mile (x)	Police Agencies per 1,000 Sq. Miles (y)
Georgia	7.30	9.40
Iowa	1.40	7.20
Kansas	1.30	4.40
Maine	1.00	3.90
Minnesota	1.90	5.30
Missouri	3.50	8.40
Nebraska	.70	3.20
South Carolina	6.90	8.40
Washington	3.80	3.70
N = 9		

10. In police agencies—much like in other types of public and private organizations—there are concerns over disparities in pay; in particular, lower-ranking officers might feel undercompensated relative to higher-ranking command staff and administrators. Let us investigate whether there is a correlation between the pay of those at the top of the police hierarchy (chiefs) and those at the bottom (officers). We will use a random sample of six agencies from the Law Enforcement Management and Administrative Statistics data set. The table below shows the minimum annual salaries of chiefs and officers among the agencies in this sample. (The numbers represent thousands; for instance, 57.5 means $57,500.) Using an alpha level of .05, test the null hypothesis of no correlation against the alternative hypothesis that the two salary types are correlated. Use all five steps. If appropriate, interpret the sign, magnitude, and coefficient of determination.

Agency	Minimum Chief Salary (x)	Minimum Officer Salary (y)
A	57.50	28.40
B	20.40	14.50
C	57.60	31.60
D	70.00	21.50
E	21.70	16.50
F	75.50	33.80
N = 6		

11. It is well known that handguns account for a substantial portion of murders. This has led some people to claim that stricter handgun regulations would help curb the murder rate in the U.S. Others, though, say that tougher gun laws would not work because people who are motivated to kill but who cannot obtain a handgun will simply find a different weapon instead. This is called a *substitution effect*. If the substitution effect is operative, then there should be a negative correlation between handgun and knife murders. The table below contains data from a random sample of states. For each state, the handgun and knife murder rates (per 100,000 state residents) are shown. Using an alpha level of .01, test the null hypothesis that the variables are not correlated against the alternative hypothesis that they are negatively correlated. Use all five steps. If appropriate, interpret the sign, magnitude, and coefficient of determination.

State	Handgun Murder Rate (x)	Knife Murder Rate (y)
Connecticut	1.45	.48
Indiana	2.12	.53
Missouri	2.84	.67
Nebraska	1.22	.45
Wisconsin	1.15	.39
Arkansas	1.87	.73
North Carolina	2.59	.52
Virginia	1.37	.52
California	2.77	.79
Washington	1.13	.53
$N = 10$		

12. Does it take defendants who face multiple charges longer to get through the adjudication process? The table below contains data on a random sample of juveniles from the Juvenile Defendants in Criminal Courts data set. The variables are *number of charges* and *months from anchor to adjudication*, the latter of which measures the total amount of time that it took for these juveniles to have their cases disposed of. Using an alpha level of .05, test the null hypothesis that the variables are not correlated against the alternative hypothesis that they are correlated. Use all five steps. If appropriate, interpret the sign, magnitude, and coefficient of determination.

Juvenile	Charges (x)	Months (y)
A	5	7.13
B	3	5.50
C	3	2.47

Juvenile	Charges (x)	Months (y)
D	2	1.20
E	4	.47
F	3	.80
G	2	.07
H	1	.00
I	1	.93
J	1	.67
N = 10		

13. Advocates of the death penalty argue that capital punishment deters homicide by showing would-be murderers what will happen to them if they willfully take a life. From this perspective, then, the number of people executed in a state should be negatively correlated with homicide rates. Let us test this prediction using capital punishment data from BJS and homicide data from the UCR. The data set is *Executions and Homicide for Chapter 13.sav* (**http://www.sagepub.com/gau**). The independent variable is *executions* (the number of people executed in each of the 37 states that authorized the death penalty in 2008) and the DV is *homicide* (the homicide rate per 100,000 in 2009). Do the following:

   a. Run an analysis to test for a negative correlation between these two variables.
   b. Identify the obtained value of the correlation coefficient *r*.
   c. Would you reject the null hypothesis at an alpha level of .05? Why or why not?
   d. State your substantive conclusion about whether or not capital punishment is negatively correlated with homicide.
   e. If appropriate, interpret the sign, magnitude, and coefficient of determination.

14. When prisons have too few correctional officers relative to the number of inmates those officers are in charge of, the security of the institution may be threatened. One of the consequences might be a greater number of inmate escapes because there are not enough staff members to effectively supervise prisoners. The data set *Security and Escapes for Chapter 13.sav* at (**http://www.sagepub.com/gau**) contains data from the Census of State and Federal Correctional Facilities. The IV (*ratio*) is each facility's inmate-to-officer ratio; higher numbers mean more inmates per security officer, which represents an enhanced threat to control within the facility. The DV (*escapes*) is the number of escapes reported by each facility per 1,000 inmates. Do the following:

   a. Run an analysis to test for a positive correlation between these two variables.
   b. Identify the obtained value of the correlation coefficient *r*.
   c. Would you reject the null hypothesis at an alpha level of .01? Why or why not?
   d. State your substantive conclusion about whether or not staffing is positively correlated with inmate escapes.
   e. If appropriate, interpret the sign, magnitude, and coefficient of determination.

KEY TERMS

Pearson's correlation

Positive correlation

Negative correlation

Linear relationship

r coefficient

GLOSSARY OF SYMBOLS AND ABBREVIATIONS INTRODUCED IN THIS CHAPTER

r	The correlation coefficient for a sample
ρ	The correlation coefficient for a population
r^2	The coefficient of determination

REFERENCES

Bureau of Justice Statistics. (2004). Survey *of inmates in state and federal correctional facilities, 2004.* Ann Arbor, MI: Inter-University Consortium for Political and Social Research.

Henson, B., Reyns, B. W., Klahm, C. F., IV, & Frank, J. (2010). Do good recruits make good cops? Problems predicting and measuring academy and street-level success. *Police Quarterly, 13*(1), 5–26.

Morgan, K. O., Morgan, S., & Boba, R. (2010). *Crime: State rankings 2010.* Washington, DC: CQ Press.

Reisig, M. D., Pratt, T. C., & Holtfreter, K. (2009). Perceived risk of Internet theft victimization: Examining the effects of social vulnerability and financial impulsivity. *Criminal Justice & Behavior, 36*(4), 369–384.

Chapter 14

Introduction to Regression Analysis

I n Chapter 13, you learned about correlation analyses. Correlation is a statistical procedure for finding two pieces of information: (1) whether or not two continuous variables are statistically related in a linear fashion; and (2) the strength of that relationship. In most criminal justice/ criminology research, however, merely finding out whether or not two variables are correlated is not sufficient. Researchers want, instead, to find out whether one of the variables (the independent variable, IV) can be used to *predict* the other one (the dependent variable, DV). **Regression analysis** does this. Regression goes a step beyond correlation by allowing researchers to determine the extent to which the IV predicts the DV.

This chapter will discuss two types of regression analyses. **Bivariate regression** employs one independent variable and one dependent variable. It is similar to bivariate correlation in many respects. **Multiple regression** uses several IVs to predict the DV. The specific type of regression modeling discussed here is **ordinary least squares (OLS) regression**. This is a fundamental form of regression modeling that is used frequently in criminology/criminal justice research. There are other types of regression, but OLS is the default procedure that is generally employed unless there is good reason for departing from it and using a different technique instead. OLS is the starting point for understanding regression, and so it is the technique covered here.

Bivariate regression: A regression analysis that uses one independent and one dependent variable.

Multiple regression: A regression analysis that uses two or more independent variables and one dependent variable.

Ordinary least squares regression: A common procedure for estimating regression equations.

▣ ONE INDEPENDENT VARIABLE AND ONE DEPENDENT VARIABLE: BIVARIATE REGRESSION

Bivariate regression is an extension of bivariate correlations; therefore, we will use the data from Example 2 in Chapter 13 as an introduction to regression. Recall that this analysis involved the relationship between the age of onset of criminal offending (measured as the age at which an offender experienced her first arrest) and offending over the life course (measured as each inmate's total number of lifetime arrests). The sample was from the Survey of Inmates in Federal Correctional Facilities (SIFCF; Data Sources 13.1) and consisted of female inmates incarcerated for violent offenses. The data from Table 13.3 are reproduced in Table 14.1. They are sorted in ascending order according to age of onset (the IV).

Table 14.1   Age at First Arrest and Total Number of Arrests Over Life in a Sample of Female Federal Inmates Incarcerated for Violent Offenses

Person	Age at First Arrest (x)	Lifetime Arrests (y)
J	10	6
R	12	21
S	12	21
V	13	61
E	15	11
M	15	2
O	15	11
I	18	1
L	18	4
T	19	4
G	21	21
N	26	2
X	26	3
P	27	6
Y	27	3
A	28	4
W	28	2
Q	29	2
C	31	3
B	33	3
H	40	3
U	42	2
D	43	2
F	47	3
K	66	2
$N = 25$		

Every x score in Table 14.1 has a corresponding y score. These scores form the graphing coordinates (x, y). These coordinates can be graphed on a scatterplot like that in Figure 14.1. You can find each person on the graph according to that person's x and y scores. For example, Person J was 10 when she was first arrested ($x = 10$) and she reported having been arrested six times ($y = 6$). Person J thus has the coordinates (10, 6). Can you locate her on the scatterplot?

We already know that these two variables are correlated, but now we want to know how well age of onset *predicts* lifetime arrests. In other words, we want to know the extent to which knowing a woman's age of onset helps us predict the number of times she has been arrested. Correlation does not allow us to do this, but regression does.

Think about drawing a line in the above scatterplot. You want the line to come as close to as many of the data points as possible. What would this line look like? Would it have a positive or negative slope? Where would it cross the y (vertical) axis? How steep would it be? Answering these questions will inform us as to how well x predicts y.

Figure 14.1   Scatterplot of Age of Onset and Lifetime Arrests

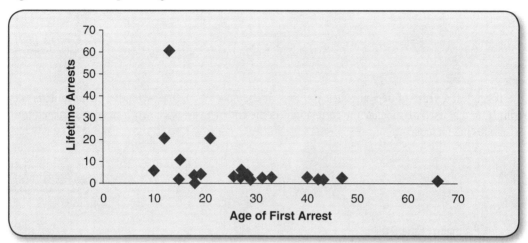

The line of best fit is the line that comes closer to each of the data points than would any other line that could possibly be drawn. The line of best fit is given by the formula:

$$\hat{Y} = a + bx, \text{where}$$

Formula 14(1)

$\hat{Y}$ = the predicted value of y at x,

a = the y-intercept,

b = the slope coefficient,

x = the raw values of the independent variable.

The predicted values of y—symbolized $\hat{Y}$ (pronounced "y hat")—are made up of the intercept a and a certain slope coefficient b that is constant across all values of x. The **intercept** is the point at which the line of best fit crosses the y-axis, while the **slope** coefficient conveys information about the steepness of that line.

Once the line given by Formula 14.1 is estimated, you have two different sets of DV scores: the actual, empirical scores (these are the y values); and the predicted values (the $\hat{Y}$ scores). If the IV is a good predictor of the DV, these two sets of numbers will be very similar to one another. If the IV does not predict the DV very well, the predicted values will differ substantially from the empirical ones.

Intercept: The point at which the regression line crosses the y-axis.

Slope: The steepness of the regression line and a measure of the change in the dependent variable produced by a one-unit increase in an independent variable.

To construct the regression line, b and a must be calculated. The slope coefficient b is calculated as

$$b = \frac{N\sum xy - \sum x \sum y}{N\sum x^2 - (\sum x)^2}.$$

Formula 14(2)

If you feel a sense of déjà vu when you look at Formula 14(2), that is good! This formula is very similar to that used to calculate the correlation coefficient r. Once b is known, a can be calculated from the following formula:

$$a = \bar{y} - b\bar{x} \text{, where}$$

Formula 14(3)

$\bar{y}$ = the mean of the DV,

b = the slope coefficient,

$\bar{x}$ = the mean of the IV.

Now we can construct the regression equation for our data. The first step is to calculate the slope coefficient b. In Chapter 13, Example 2, we computed the sums required by Formula 14(2), so these numbers can be pulled from Table 13.3 and entered into the formula:

$$b = \frac{25(3678) - (661)(203)}{25(21729) - 661^2} = \frac{91950 - 134183}{543225 - 436921} = \frac{-42233}{106304} = -.40.$$

The slope b is $-.40$, meaning that for every one-year increase in offenders' age of first arrest, their total number of lifetime arrests decreases by .40, or just under one half of one arrest. Now let us find a. We first need the DV and IV means. These are found using the mean formula with which you are familiar:

$$\bar{y} = \frac{\sum y}{N} = \frac{203}{25} = 8.12$$

$$\bar{x} = \frac{\sum x}{N} = \frac{661}{25} = 26.44.$$

Plugging the means and b into Formula 14(3) yields

$$a = 8.12 - (-.40)26.44 = 8.12 - (-10.58) = 18.70.$$

Now the entire regression equation can be constructed using the pieces we just computed:

$$\hat{Y} = 18.70 + (-.40)x = 18.70 - .40x.$$

This equation can be used to predict each person's lifetime arrests on the basis of her age at the time she was first arrested. This is accomplished by entering a given person's age of onset into the equation and solving for $\hat{Y}$. Let us revisit Person J and compute her predicted number of lifetime arrests. She was 10 ($x = 10$) years old when first arrested, so

$$\hat{Y}_{Person\,J} = 18.70 - .40(10) = 18.70 - 4.00 = 14.70.$$

Person J (or any other female inmate imprisoned for a violent offense who was 10 at the time of her first arrest) is predicted to have experienced 14.70 lifetime arrests. It is this predictive capacity that gives regression such an edge over correlation.

The regression equation can even be used to predict the number of lifetime arrests for people who are not in the sample. Nobody in the sample, for instance, reported having been 14 years old when first arrested, but this value can nevertheless be plugged into the equation to derive a predicted score:

$$\hat{Y}_{14yrs.} = 18.70 - .40(14) = 18.70 - 5.60 = 13.10.$$

A person who was 14 when first arrested is predicted to have 13.10 lifetime arrests. We know this despite the fact that nobody in this sample was actually 14 at the time of her initial arrest—that is the great thing about regression!

Inferential Regression Analysis: Testing for the Significance of *b*

The most common use of regression analysis in criminal justice/criminology research is in the context of hypothesis testing. Just like the correlation coefficient *r*, the slope coefficient *b* does not itself determine whether or not the null hypothesis should be rejected. To make this determination, a five-step hypothesis test must be conducted. We will conduct this test using $\alpha = .05$.

Step 1: State the Null and Alternative Hypotheses

The null hypothesis in regression is generally that there is no relationship between the IV and DV and, therefore, that the slope coefficient is zero. The alternative hypothesis is usually two-tailed; one-tailed tests are used only when there is a compelling reason to do so. Here, we will use a two-tailed test because this is the more customary course of action. The null and alternative hypotheses are, as always, phrased in terms of the population parameters. In regression, *B* symbolizes the population slope coefficient. The hypotheses are

$$H_0: B = 0$$

$$H_1: B \neq 0.$$

Step 2: Identify the Distribution and Compute the Degrees of Freedom

The t distribution is the one typically used in regression. When the sample size is large, z can be used instead; however, since t can accommodate any sample size, it is more efficient to simply use that distribution in all circumstances. In bivariate regression, the degrees of freedom are calculated as

$$df = N - 2. \qquad \text{Formula 14(4)}$$

Here,

$$df = 25 - 2 = 23.$$

Step 3: Identify the Critical Value and State the Decision Rule

With a two-tailed test, $\alpha = .05$, and $df = 23$, $t_{crit} = \pm 2.069$. The decision rule states: *If t_{obt} is either $<$ -2.069 or > 2.069, H_0 will be rejected.*

Step 4: Compute the Obtained Value of the Test Statistic

The first portion of this step entails calculating b and a in order to construct the regression equation in Formula 14(1). We have already done this; recall that a $= 18.70$, $b = .40$, and the equation is $\hat{y} = 18.70 - .40x$. What we have not yet done is find out whether b is statistically significant. Just like all other statistics, b has a sampling distribution. See Figure 14.2. The distribution centers on zero because the null predicts that the variables are not related. $B = 0$ is therefore the default assumption, and what we are looking for in an inferential test is evidence to lead us to believe that b is insignificantly different from zero.

Finding out whether b is statistically significant is a two-step process. First, we have to compute this coefficient's standard error, symbolized SE_b. The standard error is the standard deviation of the sampling distribution depicted in Figure 14.1. The standard error is important because all else being equal, slope coefficients with larger standard errors are less trustworthy than those with smaller standard errors. A large standard error means that there is substantial uncertainty as to the accuracy of the sample slope coefficient b as an estimate of the population slope B. SE_b is computed as:

$$SE_b = \frac{s_y}{s_x}\sqrt{\frac{1 - r^2}{N - 2}}, \text{where} \qquad \text{Formula 14(5)}$$

s_x = the standard deviation of x (the IV),

s_y = the standard deviation of y (the DV),

r = the correlation between x and y.

Since you already know how to compute standard deviations, they will simply be provided here rather than the entire process being shown. The standard deviation of y (the DV, here *number of arrests*) is 12.65 and the standard deviation of x (the IV, here *age of first arrest*) is 13.31. From Chapter 13, we know that $r = -.42$. Plugging all of these numbers in to Formula 14(5) yields

$$SE_b = \frac{12.65}{13.31}\sqrt{\frac{1-(-.42)^2}{25-2}} = .95\sqrt{\frac{1-.18}{23}} = .95\sqrt{.04} = .95(.20) = .19.$$

Now SE_b can be entered into the t_{obt} formula, which is

$$t_{obt} = \frac{b}{SE_b}.$$

Formula 14(6)

The obtained value of t is the ratio between the slope coefficient and its standard error. Entering our numbers into the equation results in

$$t_{obt} = \frac{-.40}{.19} = -2.11.$$

Step 4 is complete! The obtained value of t is -2.11. We can now make a decision about the statistical significance of b.

Figure 14.2   The Sampling Distribution of Slope Coeffecients

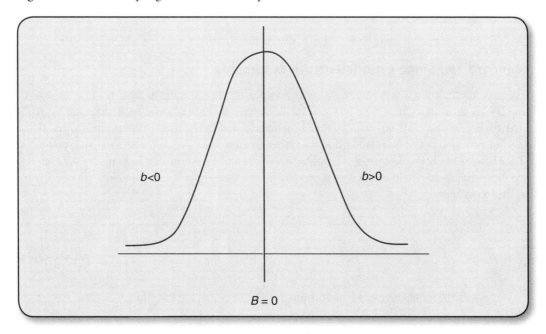

Step 5: Make a Decision About the Null and State the Substantive Conclusion

The decision rule stated that the null would be rejected if t_{obt} ended up being either less than -2.069 or greater than 2.069. Since -2.11 is less than -2.069, the null is rejected. The slope is statistically significant at an alpha of .05. There is a negative relationship between female inmates' age of first arrest and their total number of lifetime arrests; more specifically, age of first arrest helps predict total lifetime arrests.

As with correlation, rejecting the null requires further examination of the IV-DV relationship to determine the strength and quality of that connection. In the context of regression, a rejected null indicates that the IV exerts some level of predictive power over the DV; however, it is desirable to know the magnitude of this predictive capability. The following section describes two techniques for making this assessment.

Beyond Statistical Significance: How well does the independent variable perform as a predictor of the dependent variable?

There are two ways to assess model quality. The first is to create a standardized slope coefficient or **beta weight** so the slope coefficient's magnitude can be gauged. The second is to examine the coefficient of determination. Each will be discussed in turn below. Remember that these techniques should only be used when the null hypothesis has been rejected—if the null is retained, the analysis stops because the conclusion is that there is no relationship between the independent and dependent variables.

Beta weight: A standardized slope coefficient that ranges from −1.00 to +1.00 and can be interpreted similarly to a correlation so that the magnitude of an IV-DV relationship can be assessed.

Standardized Slope Coefficients: Beta Weights

The slope coefficient b is unstandardized, which means that it is specific to the units in which the IV and DV are measured. There is no way to "eyeball" an unstandardized slope coefficient and assess its strength because there are no boundaries or benchmarks that can be used with unstandardized statistics. The way to solve this is to standardize b and thereby create what is called a beta weight (symbolized β, the Greek letter beta). Beta weights range between 0.00 and ±1.00 and can be evaluated according to the same criteria used to interpret correlation coefficients (see Figure 13.2). Standardization is accomplished as follows:

$$\beta = b\left(\frac{s_x}{s_y}\right).$$

Formula 14(7)

We have all the numbers needed for Formula 14(7), so we can plug them in and solve:

$$\beta = -.40\left(\frac{13.31}{12.65}\right) = -.40(1.05) = -.42.$$

The beta weight is −.42. If this number seems familiar, it is! The correlation between these two variables is also −.42. Beta weights will equal correlations in the bivariate context and can be interpreted the same way. A beta of −.42 is fairly strong. Age of first arrest appears to be robustly associated with lifetime arrests.

The Quality of Prediction: The Coefficient of Determination

Beta weights help assess the magnitude of the relationship between an IV and a DV, but they do not provide information about how well the IV performs at actually predicting the DV. This is a substantial limitation because prediction is the heart of regression—it is the reason researchers use this technique. The coefficient of determination addresses the issue of the quality of prediction. It does this by comparing the actual, empirical values of y to the predicted values ($\hat{Y}$). A close match between these two sets of scores indicates that x does a good job predicting y, while a poor correspondence signals that x is not a useful predictor. The coefficient of determination is given by

$$\text{coefficient of determination} = r_{y\hat{y}}^2, \text{where} \qquad \boxed{\textit{Formula 14(8)}}$$

$r_{y\hat{y}}$ = the correlation between the actual and predicted values of y.

The correlation between the y and $\hat{y}$ values is computed the same way that correlations between IVs and DVs are and so will not be shown here. The correlation in this example is −.42. This makes the coefficient of determination

$$(-.42)^2 = .18.$$

This means that 18% of the variance in y can be attributed to the influence of x. In the context of the present example, 18% of the variance in lifetime arrests can be traced to the age at which a person was arrested for the first time.

▣ ADDING MORE INDEPENDENT VARIABLES: MULTIPLE REGRESSION

The problem with bivariate regression—indeed, with all bivariate hypothesis tests—is that social phenomena are usually the product of many factors, not just one. Bivariate analyses raise the specter of the omitted variable bias that was discussed in Chapter 2. The use of only one IV virtually guarantees that important predictors have been erroneously excluded and that the results of the analysis are therefore suspect.

Multiple regression is the answer to this problem. Multiple regression is an extension of bivariate regression and takes the form

$$\hat{Y} = a + b_1 x_1 + b_2 x_2 + \ldots + b_k x_k. \qquad \boxed{\textit{Formula 14(9)}}$$

Revisit Formula 14(1) and compare it to 14(9) to see how 14(9) expands upon the bivariate equation by including multiple independent variables instead of just one. The subscripts show that each IV has its own slope coefficient. With k IVs in a given study, $\hat{Y}$ is the sum of each $b_k x_k$ term and the y-intercept.

In multiple regression, the relationship between each IV and the DV is assessed while controlling for the effect of the other IV or IVs. The slope coefficients in multiple regression are called **partial slope coefficients** because for each one, the relationship between the other IVs and the DV has been removed so that each partial slope represents the "pure" relationship between an IV and the DV. The ability to incorporate multiple predictors and to assess each one's unique contribution to $\hat{Y}$ is what makes multiple regression so useful.

Partial slope coefficient: A slope coefficient that measures the individual impact of an independent variable on a dependent variable while holding other independent variables constant.

The formulas involved in multiple regression are complex and are rarely used in the typical criminal justice/criminology research setting because of the prevalence of statistical software. We will, therefore, dispense with the hand calculations for this topic and instead turn to SPSS. We will continue to use the age of first arrest and lifetime arrest data from the sample of female inmates incarcerated in federal prisons.

The bivariate regression between age of first arrest and total lifetime arrests was statistically significant, indicating that these two variables are related. What happens, though, when we add another IV that might influence lifetime arrests? One factor that has been consistently been shown to be a predictor of offending is education; all else being equal, lower education is a risk factor for offending and higher education is a protective factor. Let us, then, introduce *education* into the equation. This variable is measured as the number of years of schooling that each person in the sample reported having completed. The goal is to find out (1) whether educational attainment is related to lifetime arrests, and (2) if age of onset remains statistically significant once education is controlled for.

◉ ORDINARY LEAST SQUARES REGRESSION IN SPSS

Before launching the analysis, we should revisit the null and alternative hypotheses. In multiple regression, the null and alternative each apply to every independent variable. For each IV, the null predicts that the population slope coefficient B_k is zero and the alternative predicts that it is significantly different from zero. Since the analysis in the current example has two independent variables, the null and alternative are

$$H_0: B_1 \text{ and } B_2 = 0$$

$$H_1: B_1 \text{ and/or } B_2 \neq 0.$$

Since each IV has its own null, it is possible for the null to be rejected for one of the variables and not for the other.

To run a regression analysis in SPSS, go to *Analyze* → *Regression* → *Linear*. This will produce the dialogue box shown in Figure 14.3. Move the DV and IVs into their proper locations in the right-hand spaces and then press *OK*. This will produce an output window containing the elements displayed in the following figures.

Figure 14.3 Running a Multiple Regression Analysis in SPSS

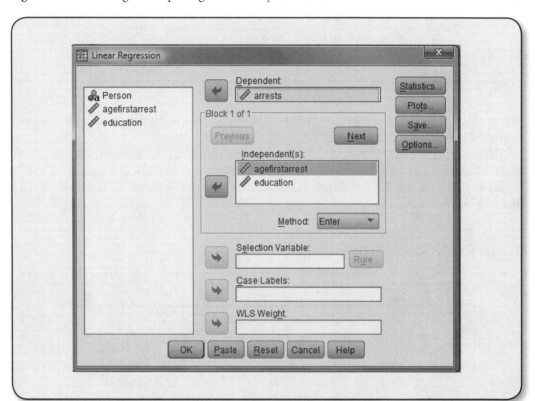

The first portion of regression output you should look at is the ANOVA box. This may sound odd since we are running a multiple regression analysis—not an ANOVA— but what this box tells you is whether or not the set of IVs included in the model explains a statistically significant amount of the variance in the DV. If *F* is not significant (i.e., if $p > .05$), then the model is no good. In the event of a nonsignificant *F*, you should *not* go on to interpret and assess the remainder of the model. Your analysis is over at that point and what you must do is revisit your data, your hypothesis, or both to find out what went wrong.

In Figure 14.4, you can see that $p = .010$, so the amount of the variance in the DV variance that is explained by the IVs is significantly greater than zero. Note that a significant *F* is not by itself proof that the model is good—this is a necessary but insufficient condition for a high-quality regression model.

Figure 14.4   SPSS Regression Output

Model		Sum of Squares	df	Mean Square	F	Sig.
1	Regression	1305.550	2	652.775	5.803	.010[a]
	Residual	2362.283	21	112.490		
	Total	3667.833	23			

ANOVA[b]

a. Predictors: (Constant), Number of years of school completed, Age at first arrest

b. Dependent Variable: Total number of lifetime arrests

What a significant F does is allow you to proceed to an examination of the slope coefficients. This portion of the output is depicted in Figure 14.4.

The first thing to look at in Figure 14.5 is the B column containing the unstandardized slope coefficients. These are the multiple regression analogs of the bivariate slope coefficient. You can see that the slope for age is $-.230$ and that for education is -1.836. The "constant" in the output is the y-intercept, a. Here, the intercept is 35.383.

Figure 14.5   SPSS Regression Output

Coefficients[a]

Model		Unstandardized Coefficients		Standardized Coefficients		
		B	Std. Error	Beta	t	Sig.
1	(Constant)	35.383	8.508		4.159	.000
	Age at first arrest	-.230	.176	-.242	-1.312	.204
	Number of years of school completed	-1.836	.710	-.476	-2.584	.017

a. Dependent Variable: Total number of lifetime arrests

The second piece of information you want is the significance level (p value) of each coefficient. Alpha was set at .05, so the null with respect to the education coefficient is rejected but that with respect to the age coefficient is retained. This is interesting! In the bivariate context, age of first arrest was significantly related to lifetime arrests; however, controlling for education washes out the effect of age, and education emerges as a significant predictor of lifetime arrests. This is an illustration of the omitted variable bias in action—had we not considered the potential effect of education, we may have erroneously concluded that age of first arrest is a major driving force behind lifetime arrests. This underscores how careful you have to be both as someone who conducts statistical analyses and as someone who is exposed to the results of other people's analyses.

RESEARCH EXAMPLE 14.1

Does having a close black friend reduce whites' concerns about crime?

In Chapter 1, you read about a study by Mears, Mancini, and Stewart (2009) in which the researchers sought to uncover whether whites' concerns about crime as a local and as a national problem were affected by whether or not those whites had at least one close friend who was black. Concern about crime was the DV in this study. White respondents expressed their attitudes about crime on a 4-point scale where higher values indicated greater concern. The researchers ran an OLS regression model and arrived at the following results with respect to whites' concerns about local crime.

IV	DV: Whites' Concern About Local Crime		
	b	*SE$_b$*	β
Have a black friend	.15*	.07	.08
Age	.06**	.02	.10
Sex	−.05	.06	−.03
Education	−.01	.03	−.01
Income	−.06*	.03	−.08
Urban	.48***	.09	.20
Intercept	1.80***	.13	
R^2= .256			

Source: Adapted from Table 2 in Mears, Mancini, and Stewart (2009).

* $p < .05$;** $p < .01$; *** $p < .001$.

The authors found, contrary to what they had hypothesized, that having a close friend who was black actually *increased* whites' concerns about crime. You can see this in the fact that the slope coefficient for *have a black friend* is statistically significant ($p < .05$) and positive. Age was also related to concern, with older respondents expressing more worry about local crime. Income had a negative slope coefficient such that a one-unit increase in annual income was associated with a .06 reduction in concern. Finally, living in an urban area substantially heightened whites' worry about crime—looking at the beta weights, you can see that *urban* is the strongest IV in the model. The researchers concluded that for whites living in urban areas, having black friends may make the crime problem seem more real and immediate because they are exposed to it through these friends.

The third consideration is the magnitude of the education slope coefficient. Since the unstandardized slope for education was significant, we can examine its accompanying beta weight, which is located in the *Beta* column. Education's beta weight is $-.476$, which represents a fairly hefty impact. It appears that education is a robust predictor of lifetime arrests, with a tendency for those with greater educational attainment to experience fewer arrests and those with lower educational achievement to experience more arrests.

Finally, the overall model is assessed using the **multiple coefficient of determination**, symbolized R^2. Figure 14.5 shows this portion of the output.

Multiple coefficient of determination: The proportion of variance in a dependent variable that is attributable to the impact of two or more independent variables operating jointly.

Figure 14.6   SPSS Regression Output

Model Summary

Model	R	R Square	Adjusted R Square	Std. Error of the Estimate
1	.597[a]	.356	.295	10.606

a. Predictors: (Constant), Number of years of school completed, Age at first arrest

The multiple coefficient of determination, R^2, is similar to the bivariate coefficient of determination, r^2. The difference is that R^2 measures the impact of several IVs rather than just one. In Figure 14.6, you can see that $R^2 = .356$, which means that these two IVs together account for 35.6% of the variance in lifetime arrests. This is respectable, as it represents over one third of the total variance in the DV. Education is a robust predictor of lifetime arrests. The adjusted R^2 is used when the purpose of the analysis is to compare the relative quality of several different models. This analytic technique is outside the scope of this book. For present purposes, the R^2 is the relevant measure of variance explained.

Recall that a major benefit of regression is that it can be used to predict values of the DV. Let us see how this works with multiple regression. Using Formula 14(9) and the numbers from the SPSS output, the regression equation can be written as

$$\hat{Y} = 35.38 + (-.23)x_{age} + (-1.84)x_{education} = 35.38 - .23x_{age} - 1.84x_{education}.$$

Suppose we want to predict the number of lifetime arrests for someone who was 14 at the time of her first arrest and completed 11 years of schooling. We would substitute these numbers into the equation and solve:

$$\hat{Y} = 35.38 - .23(14) - 1.84(11) = 35.38 - 3.22 - 20.24 = 11.92.$$

We would expect that someone who experienced her first arrest at age 14 and who achieved 11 years of education would be arrested 11.92 times during her life.

RESEARCH EXAMPLE 14.2

Do multiple homicide offenders specialize in killing?

In Chapter 11, you read about a study by Wright, Pratt, and DeLisi (2008) wherein the researchers examined whether multiple homicide offenders (MHOs) differed significantly from single homicide offenders (SHOs) in terms of the diversity of offending. Diversity was measured as a continuous variable with higher values indicated a greater spectrum of offending. We saw in Chapter 11 that Wright et al. first ran a *t* test to check for differences between the group means for MHOs and SHOs; that test showed that the difference was not statistically significant. Given that bivariate results are untrustworthy for the reasons discussed in this chapter, Wright et al. ran a multiple OLS regression model. They found the following results.

IV	DV: Diversity Index		
	b	*SE_b*	β
Offender Type: SHO	.004	.163	.001
Age	.068*	.008	.355
Age of Onset	−.156*	.011	−.551
White	−.375*	.145	−.090
Constant	−1.346		
R^2= .256			

Source: Adapted from Table 2 in Wright, Pratt, and DeLisi (2008).

*$p < .05$; **$p < .01$.

Offenders' current age, the age at which they started offending, and their race were statistically significant predictors of offending diversity. As foreshadowed by the nonsignificant *t* test, the coefficient for *Offender type: SHO* was not statistically significant. A dichotomous IV like that used here, where people in the sample were divided into two groups classified as either *MHOs* or *SHOs*, is called a **dummy variable.** The slope is

(Continued)

(Continued)

interpreted just as it is with a continuous IV: It is the amount of predicted change in the DV that occurs with a one-unit increase in the IV. Here, you can see that being an SHO increased offending diversity by only .004 of a unit. This is a very trivial change and was not statistically significant. Race is also a dummy variable, as offenders in the sample were classified as either *nonwhite* or *white*. Can you interpret this slope coefficient with respect to what it means about race as a predictor of the diversity index? If you said that white offenders score significantly lower on the diversity index, you are correct.

Wright et al.'s multiple regression model confirmed that MHOs and SHOs do not differ in terms of offending diversity. This suggests that multiple homicide offenders do not specialize in killing; to the contrary, they display as much diversity as other types of homicide offenders. The theoretical implication of this finding is that the theories that have been developed to help explain violent offending may be applicable to MHOs because these people are not a special or unique group of offenders.

In a similar way, the regression equation can be used to predict the effect of a single IV on the DV. We can figure out how much a one-unit change in the IV affects the DV while holding the other IV or IVs constant. This is generally done by increasing the value of the IV of interest by one unit while holding the other IVs at their means. This procedure allows an assessment of the unique effect of a one-unit change in the IV of interest. Let us determine the effect of one-unit increase of education while holding age constant at its mean. The mean for the age variable is 27.04. We will start with 11 as the years of education completed. Entering these numbers into the equation yields

$$\hat{Y} = 35.38 - .23(27.04) - 1.84(11) = 35.38 - 6.22 - 20.24 = 8.92.$$

Someone who was at the exact mean of the age variable (that is, was 27.04 at the time of her first arrest) and completed 11 years of school is expected to have 8.92 lifetime arrests. We will now consider the impact of a one-year increase in the amount of schooling by keeping the age variable at its mean but increasing the education variable from 11 to 12:

$$\hat{Y} = 35.38 - .23(27.04) - 1.84(12) = 35.38 - 6.22 \ 22.08 = 7.08.$$

A one-unit increase in schooling produces a reduction in predicted lifetime arrests from 8.92 to 7.08, while holding age constant. (If you notice, $7.087 - 8.92 = -1.84$, the unstandardized slope coefficient for education!) This is how the regression model can be used for prediction, and how the unique impact of each IV can be assessed while controlling for the other IVs in the model.

RESEARCH EXAMPLE 14.3

Is police academy performance a predictor of effectiveness on the job?

In Chapter 13, you encountered a study by Henson, Reyns, Klahm, and Frank (2010). The researchers sought to determine whether recruits' performance while at the academy significantly influenced their later success as police officers. Henson et al. measured success in three ways: the scores new officers received on the annual evaluations conducted by those officers' supervisors; the number of complaints lodged against these new officers; and the number of commendations they earned. These three variables are the DVs in this study.

You saw in Chapter 13 that the bivariate correlations indicated mixed support for the prediction that academy performance was related to on-the-job performance; however, to fully assess this possible link, the researchers ran an OLS regression model. They obtained the following results.

	Dependent Variables		
	Evaluation	*Complaints*	*Commendations*
IV	b (SE_b)	b (SE_b)	b (SE_b)
Civil Service exam	−.07 (.02)	−.01 (.01)	.00 (.01)
Overall academy score	.06* (.03)	−.004 (.01)	−.003 (.01)
Physical agility rating	−.01 (.03)	.02 (.02)	.00 (.02)
Gender	−.40 (.25)	−.32* (.12)	−.07 (.14)
Age	−.04 (.02)	−.02* (.01)	.01 (.01)
Race	.13 (.23)	−.10 (.11)	.14 (.13)

(Continued)

(Continued)

	Dependent Variables		
	Evaluation	Complaints	Commendations
IV	**b** (SE$_b$)	**b** (SE$_b$)	**b** (SE$_b$)
Education	−.02 (.29)	−.15 (.14)	.14 (.16)
	R^2 = .048	R^2 = .098	R^2 = .000

Source: Adapted from Table 5 in Henson, Reyns, Klahm, and Frank (2010).

*p < .01.

The results from the three OLS models showed that recruits' civil service exam scores, physical agility exam scores, or overall academy ratings were—with only one exception—unrelated to on-the-job performance. The exception was the positive slope coefficient between overall academy ratings and evaluation scores ($b = .06$; $p < .01$). The demographic variables gender, age, race, and education also bore limited and inconsistent relationships with the three performance measures. These results seem to indicate that the types of information and training that recruits receive is not as clearly and directly related to on-the-job performance as would be ideal. There may be a need for police agencies to revisit their academy procedures to ensure that recruits are receiving training that is current, realistic, and practical in the context in which these recruits will be working once they are out on the street.

CHAPTER SUMMARY

This chapter introduced you to the basics of bivariate and multiple regression analysis. Bivariate regression is an extension of bivariate correlation and is useful because correlation allows only for a determination of the association between two variables, but bivariate regression permits an examination of how well the IV acts as a predictor of the DV. When an IV emerges as a statistically significant predictor of the DV, the IV can then be assessed for magnitude using the standardized beta weight, β, and the coefficient of determination, r^2.

The problem with bivariate analyses of all types, though, is that every phenomenon that is studied in criminal justice/criminology is the result of a combined influence of multiple factors. In bivariate analyses, it is almost certain that one or more important independent variables have been omitted. Multiple regression addresses this by allowing for the introduction of several IVs so that each one can be examined while controlling for the others' impacts. In the bivariate regression example conducted in this chapter, age of onset significantly predicted lifetime arrests; however, when education was entered into the regression model, age lost its significance and education emerged as a significant and strong predictor of lifetime arrests. This exemplifies the omitted variable bias—failing to consider the full gamut of relevant IVs can lead to erroneous results and conclusions.

This also brings us to the end of the book. You made it! You struggled at times, but you stuck with it and now you have a solid grasp on the fundamentals of criminology/criminal justice research. You know how to calculate univariate and bivariate statistics and how to conduct hypothesis tests. Just as important, you know how to evaluate the statistics and tests conducted by other people. You know to be critical, ask questions, and always be humble in arriving at conclusions because every statistical test contains some level of error, be it the Type I or Type II error rate, omitted variables, or some other source of mistake. Proceed with caution and a skeptical mind when approaching statistics as either a producer or a consumer. Make GIGO a part of your life—when the information being input into the system is deficient, the conclusions are meaningless or possibly even harmful. The bottom line: Question everything!

CHAPTER 14 REVIEW PROBLEMS

1. In Example 1 of Chapter 13, you saw how to compute a correlation between the variables *number of prior incarcerations* and *number of in-prison disciplinary reports*. The data used in this problem are reproduced below. Use the data to do the following.

Person	Incarcerations (x)	Disciplinary Reports (y)
A	14	4
B	1	3
C	2	8
D	5	1
E	2	3
N = 5		

a. Calculate the slope coefficient b.
b. Calculate the intercept a. (You must first compute the means for each variable.)
c. Write out the full regression equation.
d. Calculate the number of disciplinary reports you would expect to be received by a person with
    i.   3 prior incarcerations
    ii. 15 prior incarcerations
e. Using an alpha level of .05 and two-tailed alternative hypothesis, conduct a five-step hypothesis test to determine whether the IV is a significant predictor of the DV. Note that $s_x = 5.36$, $s_Y = 2.59$, and $r = -.09$.
f. If appropriate (that is, if you rejected the null in Part D), calculate the beta weight.

2. In the Chapter 13 review problems, you calculated a correlation between the number of crimes per square mile in a random sample of states and the number of police agencies per 1,000 square miles in those same states. The data have been reproduced below. Use the data to do the following.

State	Crime per Sq. Mile (x)	Police Agencies per 1,000 Sq. Miles (y)
Georgia	7.30	9.40
Iowa	1.40	7.20
Kansas	1.30	4.40
Maine	1.00	3.90
Minnesota	1.90	5.30
Missouri	3.50	8.40
Nebraska	.70	3.20
Pennsylvania	7.60	24.90
South Carolina	6.90	8.40
Washington	3.80	3.70
N = 10		

a. Calculate the slope coefficient b.
b. Calculate the intercept a. ($\bar{x}_x = 6.60$; $\bar{x}_y = 3.56$.)
c. Write out the full regression equation.
d. Calculate how many police agencies per 1,000 square miles you would expect in a state with
   i. 5 crimes per square mile
   ii. 10 crimes per square mile
e. Using an alpha level of .05 and two-tailed alternative hypothesis, conduct a five-step hypothesis test to determine whether the IV is a significant predictor of the DV. Note that $s_x = 2.76$, $s_Y = 6.39$, and $r = .71$.
f. If appropriate (that is, if you rejected the null in Part D), calculate the beta weight.

3. In criminal justice/criminology literature, one of the strongest and most consistent predictors of crime is socioeconomic disadvantage. Negative socioeconomic factors such as poverty and unemployment have been shown to profoundly impact crime rates. The table below contains a random sample of states. The independent variable consists of 2009 data from the U.S. Census on the percent of adults in the civilian labor force that was unemployed. The DV is UCR-derived violent crime rates per 1,000 persons. Use the data to do the following.

State	Unemployed (x)	Violent Crime Rate (y)
Alaska	8.70	6.33
Arkansas	7.30	5.18
Kentucky	7.60	2.59
Louisiana	7.60	6.20
Montana	5.60	2.54
New Jersey	6.90	3.11

State	Unemployed (x)	Violent Crime Rate (y)
Pennsylvania	6.80	3.81
Virginia	5.40	2.27
Vermont	5.60	1.31
Wyoming	4.50	2.28
N = 10		

a. Calculate the slope coefficient b.
b. Calculate the intercept a. ($\bar{x}_x = 6.60$; $\bar{x}_y = 3.56$.)
c. Write out the full regression equation.
d. Calculate many violent crimes per 1,000 citizens you would expect in a state with
   i. A 4% unemployment rate
   ii. An 8% unemployment rate
e. Using an alpha level of .05 and two-tailed alternative hypothesis, conduct a five-step hypothesis test to determine whether the IV is a significant predictor of the DV. Note that $s_x = 1.29$, $s_y = 1.76$, and $r = .79$.
f. If appropriate (that is, if you rejected the null in Part D), calculate the beta weight.

4. Deterrence theory suggests that as the number of crimes that police solve goes up, crime should decrease because would-be offenders are scared by the belief that there is a good chance that they would be caught and punished if they committed an offense. The table below contains regional data from the UCR. The independent variable is *clearance* and is the percent of violent crimes that were cleared by arrest or exceptional means in 2008. The dependent variable is *violent crime rate* and is the number of violent crimes that occurred in 2009 per 1,000 residents. Use the data to do the following:

Region	Clearance (x)	Violent Crime Rate (y)
New England	48.00	2.37
Mid-Atlantic	51.30	3.59
East North Central	35.40	3.83
West North Central	47.70	2.98
South Atlantic	47.30	4.83
East South Central	45.60	4.14
West South Central	43.70	5.32
Mountain	44.50	3.74
Pacific	44.10	3.93
N = 9		

a. Calculate the slope coefficient b.
b. Calculate the intercept a. ($\bar{x}_x = 45.29$; $\bar{x}_y = 3.86$.)

   c. Write out the full regression equation.

   d. Calculate the rate of violent crimes per 1,000 citizens you would expect in a region where

      i. 30% of violent crimes were cleared

      ii. 50% of violent crimes were cleared

   e. Using an alpha level of .05 and two-tailed alternative hypothesis, conduct a five-step hypothesis test to determine whether the IV is a significant predictor of the DV. Note that $s_x = 4.41, s_y = .88$, and $r = -.25$.

   f. If appropriate (that is, if you rejected the null in Part D), calculate the beta weight.

5. Following the theme in Question 3 above, let us now consider the possible relationship between poverty and crime. The table below contains a random sample of states and the violent crime rate DV used in Question 3, but the independent variable is now the percentage of families living below the poverty line. Use the table to do the following.

State	Poverty (x)	Violent Crime Rate (y)
Pennsylvania	8.30	3.81
Vermont	6.90	1.31
Florida	9.50	6.12
Hawaii	6.80	2.75
Iowa	7.30	2.79
Missouri	9.80	4.92
Mississippi	9.90	2.81
South Carolina	11.90	6.71
Tennessee	12.20	6.68
Texas	13.20	4.91
$N = 10$		

   a. Calculate the slope coefficient b.

   b. Calculate the intercept a. ($\bar{x}_x = 9.58$; $\bar{x}_y = 4.28$.)

   c. Write out the full regression equation.

   d. Calculate the rate of violent crimes per 1,000 citizens would you expect in a state with

      i. A 5% poverty rate

      ii. A 10% poverty rate

   e. Using an alpha level of .05 and two-tailed alternative hypothesis, conduct a five-step hypothesis test to determine whether the IV is a significant predictor of the DV. Note that $s_x = 2.29, s_y = 1.87$, and $r = .78$.

   f. If appropriate (that is, if you rejected the null in Part D), calculate the beta weight.

6. In the Chapter 13 review problems, you determined whether there was a correlation between prison dollars spent by states per capita and those states' violent crime rates.

   a. Calculate the slope coefficient b.

   b. Calculate the intercept a. ($\bar{x}_x = 1.85$; $\bar{x}_y = .56$.)

State	Handgun Murder Rate (x)	Knife Murder Rate (y)
Connecticut	1.45	.48
Indiana	2.12	.53
Missouri	2.84	.67
Nebraska	1.22	.45
Wisconsin	1.15	.39
Arkansas	1.87	.73
North Carolina	2.59	.52
Virginia	1.37	.52
California	2.77	.79
Washington	1.13	.53
$N = 10$		

c. Write out the full regression equation.

d. Calculate the rate of knife murders per 1,000 citizens would you expect in a state with

  i. A handgun murder rate of 3.00

  ii. A handgun murder rate of 1.75

e. Using an alpha level of .05 and two-tailed alternative hypothesis, conduct a five-step hypothesis test to determine whether the IV is a significant predictor of the DV. Note that $s_x = .69$, $s_Y = .13$, and $r = .70$.

f. If appropriate (that is, if you rejected the null in Part D), calculate the beta weight.

The data set Juvenile Defendants in Criminal Court for Chapter 14.sav *at* **http://www.sagepub.com/gau** *contains data from the JDCC data set. The DV* (probmonths) *is the number of months to which convicted juveniles were sentenced. The IVs are age at arrest* (age) *and number of charges* (charges)*. Use this data set to do the following.*

7. Run a bivariate regression using the DV *probmonths* and the IV *age* to test for a bivariate relationship between age and probation sentence. Do not include the other variable.

8. Report the value of the ANOVA *F* and determine whether you would reject the null at an alpha of .05.

9. Using the numbers in the output, write out the bivariate regression equation for $\hat{Y}$.

10. Run a multiple regression using both IVs. Write the multiple regression equation.

11. Using an alpha level of .05, make a decision about each of the following:

  a. Would you reject the null hypothesis with respect to *age*? Why or why not?

  b. Would you reject the null hypothesis with respect to *charges*? Why or why not?

12. Using the multiple regression equation for $\hat{Y}$, do the following.

  a. Setting *age* at its mean ($\overline{x} = 16.66$), calculate the predicted probation sentence for a juvenile facing five charges.

  b. Holding *age* constant at its mean, calculate the predicted probation sentence for a juvenile facing six charges.

  c. By how much did the predicted violent crime rate change? Was this change an increase or decrease?

13. Identify the beta weights for each of the IVs in the multiple regression equation that you identified as being statistically significant. Interpret the magnitude of these beta weights.

14. Identify the multiple coefficient of determination, R^2. Interpret R^2 as a percentage, and provide an overall assessment of the predictive capacity of these two variables.

  The data set Socioeconomics and Violence for Chapter 14.sav at **http://www.sagepub.com/gau** *contains a sample of states. Violent crime rates (the variable* violentrate*) is the DV. The IVs are: percent of households receiving food stamps or SNAP (*snap*); the percent of the adult civilian workforce that is unemployed (*unemployed*); and the percent of families that are below poverty (*poverty*). Use this data set to do the following:*

15. Run a bivariate regression using the DV *violentrate* and the IV *unemployed* to test for a relationship between unemployment and violent crime. Do not include the other variables.

16. Report the value of the ANOVA *F* and determine whether you would reject the null at an alpha of .05.

17. Using the numbers in the output, write out the regression equation for $\hat{Y}$.

18. Run a multiple regression using all three IVs. Write the multiple regression equation.

19. Using an alpha level of .05, make a decision about each of the following:

  a. Would you reject the null hypothesis with respect to *snap*? Why or why not?
  b. Would you reject the null hypothesis with respect to *unemployed*? Why or why not?
  c. Would you reject the null hypothesis with respect to *poverty*? Why or why not?

20. Using the full regression equation for $\hat{Y}$, do the following.

  a. Setting *snap* and *unemployment* at their means, calculate the predicted violent crime rate in a state with a 10% poverty rate. The mean for *snap* is 8.25, and the mean for *unemployment* is 6.77.
  b. Holding *snap* and *unemployment* constant, calculate the predicted violent crime rate in a state with an 11% poverty rate.
  c. How much did the predicted violent crime rate change, and was this change an increase or decrease?

21. Identify the beta weights for each of the IVs that you identified as being statistically significant. Interpret the magnitude of these beta weights.

22. Identify the multiple coefficient of determination, R^2. Interpret R^2 as a percentage and provide an overall assessment of the predictive capacity of these two variables.

KEY TERMS

Bivariate regression	Intercept	Partial slope coefficient
Multiple regression	Slope	Multiple coefficient of determination
Ordinary least squares regression	Beta weight	

GLOSSARY OF SYMBOLS AND ABBREVIATIONS INTRODUCED IN THIS CHAPTER

y	A given empirical value of the DV
$\hat{y}$	A given predicted value of the DV
a	The y intercept
b	In a sample, the slope of the line of best fit
B	In a population, the slope of the line of best fit
SE_b	The standard error of the slope
β	Beta weight; a standardized slope coefficient
$r_{y\hat{y}}^2$	The coefficient of determination in a bivariate regression analysis
R^2	The coefficient of determination in a multiple regression analysis

REFERENCES

Henson, B., Reyns, B. W., Klahm, C. F., IV, & Frank, J. (2010). Do good recruits make good cops? Problems predicting and measuring academy and street-level success. *Police Quarterly, 13*(1), 5–26.

Mears, D. P., Mancini, C., & Stewart, E. A. (2009). Whites' concern about crime: The effects of interracial contact. *Journal of Research in Crime and Delinquency, 46*(4), 524–552.

Wright, K. A., Pratt, T. C., & DeLisi, M. (2008).Examining offending specialization in a sample of male multiple homicide offenders. *Homicide Studies, 12*(4), 381–398.

APPENDIX A. REVIEW OF BASIC MATHEMATICAL TECHNIQUES

I n order to succeed in this class, you must have a solid understanding of basic arithmetic and algebra. This appendix is designed to help you review and brush up on your math skills.

▣ SECTION 1: DIVISION

You will be doing a lot of dividing throughout this book. The common division sign ÷ will not be used. Division will always be presented in fraction format. Instead of 6 ÷ 3, for example, you will see $\frac{6}{3}$. For example,

$$\frac{20}{4} = 5 \qquad\qquad \frac{90}{10} = 9$$

Try the following as practice.

a. $\dfrac{6}{2} =$

b. $\dfrac{14}{2} =$

c. $\dfrac{15}{3} =$

d. $\dfrac{9}{3} =$

▣ SECTION 2: MULTIPLICATION

Multiplication is another frequently used operation in this book. The common multiplication sign ×
will not be used, as the symbol x is meaningful in statistics and use of this symbol in multiplication
as well could result in confusion. The signs that will be used to designate multiplication are ( ) and •.

Also, when operands or variables are right next to one another, this is an indication that you should use multiplication. For example,

$$7(3) = 21 \qquad 7 \cdot 4 = 28 \qquad 10(3)(4) = 120$$

Try the following as practice.

  a. $3(4) =$
  b. $9 \cdot 8 =$
  c. $12(2) =$
  d. $4 \cdot 5 \cdot 3 =$

回 SECTION 3: ORDER OF OPERATIONS

Solving equations correctly requires the use of proper order of operations. The correct order is: parentheses; exponents; multiplication; division; addition; subtraction. Using any other order could result in erroneous final answers. For example,

$$3(5) + 2 = 15 + 2 = 17 \qquad (4 + 7) - 6 = 11 - 6 = 5 \qquad \left(\frac{8}{2}\right)^2 = (4)^2 = 16$$

Try the following as practice.

  a. $3 + 2 - 4 =$
  b. $4(5) + 7 =$

  c. $\dfrac{19 - 4}{5} =$

  d. $5 \cdot 6 - \dfrac{16}{4} =$

  e. $\left(\dfrac{16}{8}\right)\left(\dfrac{14}{2}\right) =$

  f. $2^2 + 3^3 =$
  g. $(3 + 2)^2 =$

回 SECTION 4: VARIABLES

The formulas in this book require you to plug numbers into equations and solve those equations. You must, therefore, understand the basic principles of algebra, wherein a formula contains variables, you are told the values of those variables, and you plug the values into the formula. For example,

$$\text{If } x = 9 \text{ and } y = 7, \text{ then } x + y = 9 + 7 = 16$$

$$\text{If } x = 10 \text{ and } y = 7, \text{ then } xy = 10(7) = 70$$

$$\text{If } x = 2, y = 5, \text{and } z = 8, \text{then } \left(\frac{z}{x}\right)y = \left(\frac{8}{2}\right)5 = 4 \cdot 5 = 20$$

Try the following as practice.

    a. $\dfrac{x}{y}$, where $x = 12$ and $y = 3$

    b. xy, where $x = 1$ and $y = 1$

    c. $x + y + z$, where $x = 1, y = 19$, and $z = 4$

    d. $\dfrac{x}{y} + 2$, where $x = 6$ and $y = 3$

    e. $\left(\dfrac{x}{6}\right)y + 5$, where $x = 36$ and $y = 11$

◉ SECTION 5: NEGATIVES

There are several rules with respect to negatives. Negative numbers and positive numbers act differently when they are added, subtracted, multiplied, and divided. Positive numbers get larger as the number line is traced away from zero and toward positive infinity. Negative numbers, by contrast, get smaller as the number line is traced toward negative infinity. Adding a negative number is equivalent to subtracting a positive number. When a positive number is multiplied or divided by a negative number, the final answer is negative. When two negative numbers are multiplied or divided, the answer is positive. For example,

$$5 + (-2) = 5 - 2 = 3 \qquad -5 + (-2) = -5 - 2 = 7 \qquad 10(9) = -90 \qquad \frac{-90}{-9} = 10$$

Try the following as practice.

    a. $-3 + (-2) =$
    b. $-3 - 4 =$
    c. $-5 + 3 =$
    d. $3 - 8 =$
    e. $(-2)^2 =$
    f. $-2^2 =$
    g. $(-4)(-5) =$

    h. $\dfrac{-9}{3} =$

◉ SECTION 6: DECIMALS AND ROUNDING

This book requires you to round. Two decimal places are used here in the text; however, your instructor may require more or fewer, so pay attention to directions. When rounding to two decimal places, you should look at the number in the third decimal place to decide whether you will round up or whether you will truncate. When the number in the third (thousandths) position is 5 or greater, you should round the number in the second (hundredths) position up. When the number in the thousandths position is 4 or less, you should truncate. The diagram below shows these positions pictorially.

x.xxx

Examples include,

.506 rounded to two decimal places = .51   .632 rounded to two decimal places = .63

$.50 + .70 = 1.20$ $(.42)(.80) = .336 \approx .34$ $\sqrt{14} = 3.742 \approx 3.74$ $\left(\dfrac{12}{9}\right)5 = (1.33)5 = 6.65$

Try the following as practice.

   a. $.50 + .55 =$

   b. $\dfrac{39}{5} =$

   c. $2.23 - .34 =$

   d. $\sqrt{12}$

   e. $1 - .66 =$

   f. $(.20)(.80) =$

   g. $\dfrac{1.90}{1.20} =$

   h. $-3\left(\dfrac{19}{8}\right) + 10 =$

   i. round this number to two decimal places: .605
   j. round this number to two decimal places: .098

   If these operations all looked familiar to you and you were able to do them with little or no difficulty, then you are ready for the course! If you struggled with them, you should speak with your course instructor regarding recommendations and options.

▣ ANSWERS TO APPENDIX A PROBLEMS

Section 1
   a.3   b.7   c.5   d.3

Section 2
   a.12   b.72   c.24   d.60

Section 3

a. 1   b. 27   c. 3   d. 36   e. 14   f. 13   g. 25

Section 4

a. 4   b. 1   c. 24   d. 4   e. 71

Section 5

a. −5   b. −7   c. −2   d. −5   e. 4   f. −4   g. 20   h. −3

Section 6

a. 1.05   b. 7.80   c. 1.89   d. 3.46   e. .34   f. .16   g. 1.58   h. 2.86   i. .61   j. .10

APPENDIX B. THE STANDARD NORMAL (z) DISTRIBUTION

🔲 **AREA BETWEEN THE MEAN AND z**

z: tenths and hundredths → z: digits and tenths ↓	.00	.01	.02	.03	.04	.05	.06	.07	.08	.09
0.0	.0000	.0040	.0080	.0120	.0160	.0199	.0239	.0279	.0319	.0359
0.1	.0398	.0438	.0478	.0517	.0557	.0596	.0636	.0675	.0714	.0753
0.2	.0793	.0832	.0871	.0910	.0948	.0987	.1026	.1064	.1103	.1141
0.3	.1179	.1217	.1255	.1293	.1331	.1368	.1406	.1443	.1480	.1517
0.4	.1554	.1591	.1628	.1664	.1700	.1736	.1772	.1808	.1844	.1879
0.5	.1915	.1950	.1985	.2019	.2054	.2088	.2123	.2157	.2190	.2224
0.6	.2257	.2291	.2324	.2357	.2389	.2422	.2454	.2486	.2517	.2549
0.7	.2580	.2611	.2642	.2673	.2704	.2734	.2764	.2794	.2823	.2852
0.8	.2881	.2910	.2939	.2967	.2995	.3023	.3051	.3078	.3106	.3133
0.9	.3159	.3186	.3212	.3238	.3264	.3289	.3315	.3340	.3365	.3389
1.0	.3413	.3438	.3461	.3485	.3508	.3531	.3554	.3577	.3599	.3621
1.1	.3643	.3665	.3686	.3708	.3729	.3749	.3770	.3790	.3810	.3830
1.2	.3849	.3869	.3888	.3907	.3925	.3944	.3962	.3980	.3997	.4015
1.3	.4032	.4049	.4066	.4082	.4099	.4115	.4131	.4147	.4162	.4177
1.4	.4192	.4207	.4222	.4236	.4251	.4265	.4279	.4292	.4306	.4319
1.5	.4332	.4345	.4357	.4370	.4382	.4394	.4406	.4418	.4429	.4441
1.6	.4452	.4463	.4474	.4484	.4495	.4505	.4515	.4525	.4535	.4545
1.7	.4554	.4564	.4573	.4582	.4591	.4599	.4608	.4616	.4625	.4633

(Continued)

(Continued)

z: tenths and hundredths → z: digits and tenths ↓	.00	.01	.02	.03	.04	.05	.06	.07	.08	.09
1.8	.4641	.4649	.4656	.4664	.4671	.4678	.4686	.4693	.4699	.4706
1.9	.4713	.4719	.4726	.4732	.4738	.4744	.4750	.4756	.4761	.4767
2.0	.4772	.4778	.4783	.4788	.4793	.4798	.4803	.4808	.4812	.4817
2.1	.4821	.4826	.4830	.4834	.4838	.4842	.4846	.4850	.4854	.4857
2.2	.4861	.4864	.4868	.4871	.4875	.4878	.4881	.4884	.4887	.4890
2.3	.4893	.4896	.4898	.4901	.4904	.4906	.4909	.4911	.4913	.4916
2.4	.4918	.4920	.4922	.4925	.4927	.4929	.4931	.4932	.4934	.4936
2.5	.4938	.4940	.4941	.4943	.4945	.4946	.4948	.4949	.4951	.4952
2.6	.4953	.4955	.4956	.4957	.4959	.4960	.4961	.4962	.4963	.4964
2.7	.4965	.4966	.4967	.4968	.4969	.4970	.4971	.4972	.4973	.4974
2.8	.4974	.4975	.4976	.4977	.4977	.4978	.4979	.4979	.4980	.4981
2.9	.4981	.4982	.4982	.4983	.4984	.4984	.4985	.4985	.4986	.4986
3.0	.4987	.4987	.4987	.4988	.4988	.4989	.4989	.4989	.4990	.4990
3.5	.4998									
4.0	.4999									

APPENDIX C. *t* DISTRIBUTION

df	Level of Significance for One-Tailed Test					
	.10	.05	.025	.01	.005	.0005
	Level of Significance for Two-Tailed Test					
	.20	.10	.05	.02	.01	.001
1	3.078	6.314	12.706	31.821	63.657	636.619
2	1.886	2.920	4.303	6.965	9.925	31.598
3	1.638	2.353	3.182	4.541	5.841	12.941
4	1.533	2.132	2.776	3.747	4.604	8.610
5	1.476	2.015	2.571	3.365	4.032	6.859
6	1.440	1.943	2.447	3.143	3.707	5.959
7	1.415	1.895	2.365	2.998	3.499	5.405
8	1.397	1.860	2.306	2.896	3.355	5.041
9	1.383	1.833	2.262	2.821	3.250	4.781
10	1.372	1.812	2.228	2.764	3.169	4.587
11	1.363	1.796	2.201	2.718	3.106	4.437
12	1.356	1.782	2.179	2.681	3.055	4.318
13	1.350	1.771	2.160	2.650	3.012	4.221
14	1.345	1.761	2.145	2.624	2.977	4.140
15	1.341	1.753	2.131	2.602	2.947	4.073
16	1.337	1.746	2.120	2.583	2.921	4.015
17	1.333	1.740	2.110	2.567	2.898	3.965
18	1.330	1.734	2.101	2.552	2.878	3.922
19	1.328	1.729	2.093	2.539	2.861	3.883
20	1.325	1.725	2.086	2.528	2.845	3.850
21	1.323	1.721	2.080	2.518	2.831	3.819
22	1.321	1.717	2.074	2.508	2.819	3.792

(Continued)

(Continued)

df	Level of Significance for One-Tailed Test					
	.10	.05	.025	.01	.005	.0005
	Level of Significance for Two-Tailed Test					
	.20	.10	.05	.02	.01	.001
23	1.319	1.714	2.069	2.500	2.807	3.767
24	1.318	1.711	2.064	2.492	2.797	3.745
25	1.316	1.708	2.060	2.485	2.787	3.725
26	1.315	1.706	2.056	2.479	2.779	3.707
27	1.314	1.703	2.052	2.473	2.771	3.690
28	1.313	1.701	2.048	2.467	2.763	3.674
29	1.311	1.699	2.045	2.462	2.756	3.659
30	1.310	1.697	2.042	2.457	2.750	3.646
40	1.303	1.684	2.021	2.423	2.704	3.551
60	1.296	1.671	2.000	2.390	2.660	3.460
120	1.289	1.658	1.980	2.358	2.617	3.373
∞	1.282	1.645	1.960	2.326	2.576	3.291

Source: Abridged from R. A. Fisher and F. Yates, *Statistical Tables for Biological, Agricultural and Medical Research*, 6th ed. Copyright © R. A. Fisher and F. Yates 1963. Reprinted by permission of Pearson Education Limited.

APPENDIX D.
CHI-SQUARE (χ^2) DISTRIBUTION

df	α			
	.10	.05	.01	.001
1	2.706	3.841	6.635	10.827
2	4.605	5.991	9.210	13.815
3	6.251	7.815	11.341	16.268
4	7.779	9.488	13.277	18.465
5	9.236	11.070	15.086	20.517
6	10.645	12.592	16.812	22.457
7	12.017	14.067	18.475	24.322
8	13.362	15.507	20.090	26.125
9	14.684	16.919	21.666	27.877
10	15.987	18.307	23.209	29.588
11	17.275	19.675	24.725	31.264
12	18.549	21.026	26.217	32.909
13	19.812	22.362	27.688	34.528
14	21.064	23.685	29.141	36.123
15	22.307	24.996	30.578	37.697
16	23.542	26.296	32.000	39.252
17	24.769	27.587	33.409	40.790
18	25.989	28.869	34.805	42.312
19	27.204	30.144	36.191	43.820
20	28.412	31.410	37.566	45.315
21	29.615	32.671	38.932	46.797
22	30.813	33.924	40.289	48.268

(Continued)

(Continued)

df	α			
	.10	*.05*	*.01*	*.001*
23	32.007	35.172	41.638	49.728
24	33.196	36.415	42.980	51.179
25	34.382	37.652	44.314	52.620
26	35.563	38.885	45.642	54.052
27	36.741	40.113	46.963	55.476
28	37.916	41.337	48.278	56.893
29	39.087	42.557	49.588	58.302
30	40.256	43.773	50.892	59.703

Source: R. A. Fisher & F. Yates, *Statistical Tables for Biological, Agricultural and Medical Research,* 6th ed. Copyright ©R. A. Fisher and F. Yates 1963. Reprinted by permission of Pearson Education Limited.

APPENDIX E. F DISTRIBUTION

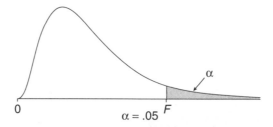

$$\alpha = .05$$

| df_W | \multicolumn{10}{c}{df_B} |
	1	2	3	4	5	6	8	12	24	∞
1	161.4	199.5	215.7	224.6	230.2	234.0	238.9	243.9	249.0	254.3
2	18.51	19.00	19.16	19.25	19.30	19.33	19.37	19.41	19.45	19.50
3	10.13	9.55	9.28	9.12	9.01	8.94	8.84	8.74	8.64	8.53
4	7.71	6.94	6.59	6.39	6.26	6.16	6.04	5.91	5.77	5.63
5	6.61	5.79	5.41	5.19	5.05	4.95	4.82	4.68	4.53	4.36
6	5.99	5.14	4.76	4.53	4.39	4.28	4.15	4.00	3.84	3.67
7	5.59	4.74	4.35	4.12	3.97	3.87	3.73	3.57	3.41	3.23
8	5.32	4.46	4.07	3.84	3.69	3.58	3.44	3.28	3.12	2.93
9	5.12	4.26	3.86	3.63	3.48	3.37	3.23	3.07	2.90	2.71
10	4.96	4.10	3.71	3.48	3.33	3.22	3.07	2.91	2.74	2.54
11	4.84	3.98	3.59	3.36	3.20	3.09	2.95	2.79	2.61	2.40
12	4.75	3.88	3.49	3.26	3.11	3.00	2.85	2.69	2.50	2.30
13	4.67	3.80	3.41	3.18	3.02	2.92	2.77	2.60	2.42	2.21
14	4.60	3.74	3.34	3.11	2.96	2.85	2.70	2.53	2.35	2.13
15	4.54	3.68	3.29	3.06	2.90	2.79	2.64	2.48	2.29	2.07
16	4.49	3.63	3.24	3.01	2.85	2.74	2.59	2.42	2.24	2.01
17	4.45	3.59	3.20	2.96	2.81	2.70	2.55	2.38	2.19	1.96
18	4.41	3.55	3.16	2.93	2.77	2.66	2.51	2.34	2.15	1.92
19	4.38	3.52	3.13	2.90	2.74	2.63	2.48	2.31	2.11	1.88
20	4.35	3.49	3.10	2.87	2.71	2.60	2.45	2.28	2.08	1.84
21	4.32	3.47	3.07	2.84	2.68	2.57	2.42	2.25	2.05	1.81
22	4.30	3.44	3.05	2.82	2.66	2.55	2.40	2.23	2.03	1.78
23	4.28	3.42	3.03	2.80	2.64	2.53	2.38	2.20	2.00	1.76
24	4.26	3.40	3.01	2.78	2.62	2.51	2.36	2.18	1.98	1.73
25	4.24	3.38	2.99	2.76	2.60	2.49	2.34	2.16	1.96	1.71
26	4.22	3.37	2.98	2.74	2.59	2.47	2.32	2.15	1.95	1.69
27	4.21	3.35	2.96	2.73	2.57	2.46	2.30	2.13	1.93	1.67
28	4.20	3.34	2.95	2.71	2.56	2.44	2.29	2.12	1.91	1.65
29	4.18	3.33	2.93	2.70	2.54	2.43	2.28	2.10	1.90	1.64
30	4.17	3.32	2.92	2.69	2.53	2.42	2.27	2.09	1.89	1.62
40	4.08	3.23	2.84	2.61	2.45	2.34	2.18	2.00	1.79	1.51
60	4.00	3.15	2.76	2.52	2.37	2.25	2.10	1.92	1.70	1.39
120	3.92	3.07	2.68	2.45	2.29	2.17	2.02	1.83	1.61	1.25
∞	3.84	2.99	2.60	2.37	2.21	2.09	1.94	1.75	1.52	1.00

Source: R. A. Fisher and F. Yates, *Statistical Tables for Biological, Agricultural and Medical Research*, 6th ed. Copyright © R. A. Fisher and F. Yates 1963. Reprinted by permission of Pearson Education Limited.

					df_B					
df_w	1	2	3	4	5	6	8	12	24	∞
1	4052	4999	5403	5625	5764	5859	5981	6106	6234	6366
2	98.49	99.01	99.17	99.25	99.30	99.33	99.36	99.42	99.46	99.50
3	34.12	30.81	29.46	28.71	28.24	27.91	27.49	27.05	26.60	26.12
4	21.20	18.00	16.69	15.98	15.52	15.21	14.80	14.37	13.93	13.46
5	16.26	13.27	12.06	11.39	10.97	10.67	10.27	9.89	9.47	9.02
6	13.74	10.92	9.78	9.15	8.75	8.47	8.10	7.72	7.31	6.88
7	12.25	9.55	8.45	7.85	7.46	7.19	6.84	6.47	6.07	5.65
8	11.26	8.65	7.59	7.01	6.63	6.37	6.03	5.67	5.28	4.86
9	10.56	8.02	6.99	6.42	6.06	5.80	5.47	5.11	4.73	4.31
10	10.04	7.56	6.55	5.99	5.64	5.39	5.06	4.71	4.33	3.91
11	9.65	7.20	6.22	5.67	5.32	5.07	4.74	4.40	4.02	3.60
12	9.33	6.93	5.95	5.41	5.06	4.82	4.50	4.16	3.78	3.36
13	9.07	6.70	5.74	5.20	4.86	4.62	4.30	3.96	3.59	3.16
14	8.86	6.51	5.56	5.03	4.69	4.46	4.14	3.80	3.43	3.00
15	8.68	6.36	5.42	4.89	4.56	4.32	4.00	3.67	3.29	2.87
16	8.53	6.23	5.29	4.77	4.44	4.20	3.89	3.55	3.18	2.75
17	8.40	6.11	5.18	4.67	4.34	4.10	3.79	3.45	3.08	2.65
18	8.28	6.01	5.09	4.58	4.25	4.01	3.71	3.37	3.00	2.57
19	8.18	5.93	5.01	4.50	4.17	3.94	3.63	3.30	2.92	2.49
20	8.10	5.85	4.94	4.43	4.10	3.87	3.56	3.23	2.86	2.42
21	8.02	5.78	4.87	4.37	4.04	3.81	3.51	3.17	2.80	2.36
22	7.94	5.72	4.82	4.31	3.99	3.76	3.45	3.12	2.75	2.31
23	7.88	5.66	4.76	4.23	3.94	3.71	3.41	3.07	2.70	2.26
24	7.82	5.61	4.72	4.22	3.90	3.67	3.36	3.03	2.66	2.21
25	7.77	5.57	4.68	4.18	3.86	3.63	3.32	2.99	2.62	2.17
26	7.72	5.53	4.64	4.14	3.82	3.59	3.29	2.96	2.58	2.13
27	7.68	5.49	4.60	4.11	3.78	3.56	3.26	2.93	2.55	2.10
28	7.64	5.45	4.57	4.07	3.75	3.53	3.23	2.90	2.52	2.06
29	7.60	5.42	4.54	4.04	3.73	3.50	3.20	2.87	2.49	2.03
30	7.56	5.39	4.51	4.02	3.70	3.47	3.17	2.84	2.47	2.01
40	7.31	5.18	4.31	3.83	3.51	3.29	2.99	2.66	2.29	1.80
60	7.08	4.98	4.13	3.65	3.34	3.12	2.82	2.50	2.12	1.60
120	6.85	4.79	3.95	3.48	3.17	2.96	2.66	2.34	1.95	1.38
∞	6.64	4.60	3.78	3.32	3.02	2.80	2.51	2.18	1.79	1.00

$\alpha = .01$

GLOSSARY

Alpha level: The opposite of the confidence level; that is, the probability that a confidence interval does not contain the true population parameter.

Alternative hypothesis: In an inferential test, the hypothesis predicting that there is a relationship between the independent and dependent variables. Symbolized H_1.

Analysis of variance (ANOVA): The analytic technique appropriate when an independent variable is categorical with three or more classes and a dependent variable is continuous.

Beta weight: A standardized slope coefficient that ranges from -1.00 to $+1.00$ and can be interpreted similarly to a correlation so that the magnitude of an IV-DV relationship can be assessed.

Between-group variance: The extent to which a set of groups or classes are similar to or different from one another. This is a measure of true group effect, or a relationship between the independent and dependent variables.

Binomial: A trial with exactly two possible outcomes. Also called a *dichotomous* or *binary variable empirical* outcome.

Bivariate: Analysis involving two variables. Usually, one is designated the independent variable and the other the dependent variable.

Bivariate regression: A regression analysis that uses one independent and one dependent variable.

Bonferroni: A widely used and relatively conservative post hoc test used in ANOVA when the null is rejected as a means of determining the number and location of differences between groups.

Bounding rule: The rule stating that all proportions range from 0.00 to 1.00.

Categorical variable: A variable that classifies people or objects into groups. There are two types: *nominal* and *ordinal*.

Cell: The place in a table where a row and column meet.

Central limit theorem: The property of the sampling distribution that guarantees that this curve will be normally distributed when infinite samples of large size have been drawn.

Chi-square test of independence: The hypothesis testing procedure appropriate when the independent and dependent variables are both categorical.

Classes: The categories or groups on a nominal or ordinal variable.

Confidence interval: A range of values spanning a point estimate that is calculated so as to have a certain probability of containing the population parameter.

Constant: A characteristic that takes on only one value in a sample or population.

Contingency table: A table showing the overlap between two variables.

Continuous Variable: A variable that numerically measures the presence of a particular characteristic. There are two types: interval and ratio.

Cramer's V: A symmetrical measure of association for χ^2 when the variables are nominal. V ranges from 0.00 to 1.00 and indicates the strength of the IV-DV relationship. Higher values represent stronger relationships.

Critical value: The value of z or t associated with a given alpha level.

Cumulative: A frequency, proportion, or percentage obtained by adding a given number to all numbers below it.

Dependent samples: Pairs of samples in which the selection of people or objects into one sample directly affected or was directly affected by the selection of people or objects into the other sample. The most common types are *matched pairs* and *repeated measures*.

Dependent variable: The phenomenon that a researcher wishes to study, explain, or predict.

Descriptive research: Studies done solely for the purpose of describing a particular phenomenon as it occurs in a sample.

Deviation score: The distance between the mean of a data set and any given raw score in that set.

Dispersion: The amount of spread or variability among the scores in a distribution.

Ecological fallacy: The error of assuming that a statistical relationship that is present in a group applies uniformly to all individual people or objects within that group.

Empirical: Having the qualities of being measurable, observable, or tangible. Empirical phenomena are detectable with senses such as sight, hearing, or touch.

Empirical outcome: A numerical result from a sample, such as a mean or frequency. Also called *observed outcomes*.

Evaluation research: Studies intended to assess the results of programs or interventions for purposes of discovering whether those programs or interventions appear to be effective.

Expected frequencies: The theoretical results that would be seen if the null were true, that is, if the IV and DV were, in fact, unrelated. Symbolized f_e.

Exploratory research: Studies that address issues that not been examined much or at all in prior research and that therefore may lack firm theoretical and empirical grounding.

F distribution: The sampling/probability distribution for ANOVA. The distribution is bounded at zero on the left and extends to positive infinity; all values in the F distribution are thus positive.

F statistic: The statistic utilized in ANOVA; a ratio of the amount of between-group variance present in a sample relative to the amount of within-group variance.

Familywise error: The increase in the likelihood of a Type I error (erroneous rejection of a true null hypothesis) that results from running repeated statistical tests on a single sample.

Frequency: A raw count of the number of times a particular characteristic appears in a data set.

Goodman and Kruskal's gamma: A symmetrical measure of association for χ^2 when the variables are ordinal. Gamma ranges from -1.00 to $+1.00$.

Groups: Classes on a categorical independent variable.

Hypothesis: A single proposition, deduced from a theory, that must hold true in order for the theory itself to be considered valid.

Independent samples: Pairs of samples in which the selection of people or objects into one sample in no way affected or was affected by the selection of people or objects into the other sample.

Independent variable: A factor or characteristic that is used to try to explain or predict a dependent variable.

Inferential analysis: The process of generalizing from a sample to a population; the use of a sample statistic to estimate a population parameter. Also called *hypothesis testing* in reference to the process of testing the validity of a theoretical concept using a sample and inferential procedures.

Inferential statistics: The field of statistics in which a descriptive statistic derived from a sample is employed probabilistically to make a generalization or inference about the population from which the sample was drawn.

Intercept: The point at which the regression line crosses the *y*-axis.

Interval variable: A quantitative variable that numerically measures the extent to which a particular characteristic is present or absent and does not have a true zero point.

Lambda: An asymmetrical measure of association for χ^2 when the variables are nominal. Lambda ranges from 0.00 to 1.00 and is a proportionate reduction in error measure.

Level of confidence: The probability that a confidence interval contains the population parameter. Commonly set at 95% or 99%.

Level of measurement: A variable's specific type or classification. There are four types: *nominal, ordinal, interval,* and *ratio.*

Linear relationship: A relationship wherein the change in the dependent variable that is produced by a one-unit increase in the independent variable remains static or constant at all levels of the independent variable.

Longitudinal variable: A variable measured repeatedly over time.

Mean: The arithmetic average of a set of data.

Measures of association: Procedures for determining the strength or magnitude of a relationship between an IV and a DV. Used only when the null hypothesis has been rejected.

Measures of central tendency: Descriptive statistics that offer information about where the scores in a particular data set tend to cluster. Examples include the *mode, median,* and *mean.*

Median: The score that cuts a distribution exactly in half such that 50% of the scores are above that value and 50% are below it.

Methods: The procedures used to gather and analyze scientific data.

Midpoint of the magnitudes: The property of the mean that causes all deviation scores based on the mean to sum to zero.

Mode: The most frequently occurring category or value in a set of scores.

Multiple coefficient of determination: The proportion of variance in a dependent variable that is attributable to the impact of two or more independent variables operating jointly.

Multiple regression: A regression analysis that uses two or more independent variables and one dependent variable.

Negative correlation: When a one-unit increase in the independent variable produces a reduction in the dependent variable.

Nominal variable: A classification that places people or objects into different groups according to a particular characteristic that cannot be ranked in terms of quantity.

Normal curve: A distribution of raw scores from a sample or population that is symmetric, unimodal, and has an area of 1.00. Normal curves differ from one another in metrics, means, and standard deviations.

Null hypothesis: In an inferential test, the hypothesis predicting that there is no relationship between the independent and dependent variables. Symbolized H_0.

Observed frequencies: The empirical results seen in a contingency table derived from sample data. Symbolized f_o.

Obtained value: The value of the test statistic arrived at using the mathematical formulas specific to a particular test. The obtained value is the final product of Step 4 of a hypothesis test.

Omega squared: A measure of association used in ANOVA when the null has been rejected in order to assess the magnitude of the relationship between the independent and dependent variables. This measure shows the proportion of the total variability in the sample that is attributable to between-group differences.

Omitted variable: An independent variable that is significantly related to a dependent variable but has been erroneously excluded from the statistical analysis.

Omitted variable bias: The erroneous conclusion that there is a relationship between an independent and dependent variable when, in fact, that relationship is explained by a third variable that has not been included in the analysis.

One-tailed tests: Hypothesis tests in which the entire alpha is placed in either the upper (positive) or lower (negative) tail such that there is only one critical value of the test statistic. Also called *directional tests*.

Ordinal variable: A classification that places people or objects into different groups according to a particular characteristic that can be ranked in terms of quantity.

Ordinary least squares regression: A common procedure for estimating regression equations.

p **value:** In SPSS output, the probability associated with the obtained value of the test statistic. When $p < \alpha$, the null hypothesis is rejected.

Partial slope coefficient: A slope coefficient that measures the individual impact of an independent variable on a dependent variable while holding other independent variables constant.

Pearson's correlation: The bivariate statistical analysis used when the independent and dependent variables are both continuous.

Percentage: A standardized form of a frequency that ranges from 0.00 to 100.00.

Point estimate: A sample statistic, such as a mean or proportion.

Pooled variances: The type of *t* test appropriate when the samples are independent and the population variances are equal.

Population: The universe of people, objects, or locations that researchers wish to study. These groups are often very large.

Population distribution: An empirical distribution made of raw scores from a population.

Positive correlation: When a one-unit increase in the independent variable produces an increase in the dependent variable.

Post hoc tests: Analyses conducted when the null is rejected in ANOVA in order to determine the number and location of differences between groups.

Probability: The likelihood that a certain event will occur.

Probability distribution: A table or graph showing the entire set of probabilities associated with every possible trial.

Probability sampling: A sampling technique in which all people, objects, or areas in a population have an equal and known chance of being selected into the sample.

Probability theory: Logical premises that form a set of predictions about the likelihood of certain events or the empirical results that one would expect to see in an infinite set of trials.

Proportion: A standardized form of a frequency that ranges from 0.00 to 1.00.

r coefficient: The test statistic in a correlation analysis.

Range: A measure of dispersion for continuous variables that is calculated by subtracting the smallest score from the largest.

Ratio variable: A quantitative variable that numerically measures the extent to which a particular characteristic is present or absent and that has a true zero point.

Replication: The repetition of a particular study that is conducted for purposes of determining whether the original study's results hold when new samples or measures are employed.

Rule of the complement: Based on the bounding rule, the rule stating that the proportion of cases that are not in a certain category can be found by subtracting the proportion that are in that category from 1.00.

Sample: A subset pulled from a population with the goal of ultimately using the people, objects, or places in the sample as a way to generalize to the population.

Sample distribution: An empirical distribution made of raw scores from a sample.

Sampling distribution: A theoretical distribution made out of an infinite number of sample statistics.

Sampling error: The uncertainty introduced into a sample statistic by the fact that any given sample is only one of an infinite number of samples that could have been drawn from the population.

Science: The process of gathering and analyzing data in a systematic and controlled way using procedures that are generally accepted by others in the discipline.

Separate variances: The type of t test appropriate when the samples are independent and the population variances are unequal.

Slope: The steepness of the regression line and a measure of the change in the dependent variable produced by a one-unit increase in an independent variable.

Somers' d: An asymmetrical measure of association for χ^2 when the variables are nominal. Somers' d ranges from -1.00 to $+1.00$.

Standard deviation: Computed as the square root of the variance, a measure of dispersion that is the mean of the deviation scores.

Standard error: The standard deviation of the sampling distribution.

Standard normal curve: A distribution of z scores. The curve is symmetric and unimodal, and has a mean of zero, a standard deviation of 1.00, and an area of 1.00.

Statistical dependence: The condition in which two variables are related to one another; that is, knowing what class persons or objects fall into on the IV helps predict which class they will fall into on the DV.

Statistical independence: The condition in which two variables are not related to one another; that is, knowing what class persons or objects fall into on the IV does not help predict which class they will fall into on the DV.

t **distribution:** A family of curves whose shapes are determined by the size of the sample. All t curves are unimodal, symmetrical, and have an area of 1.00.

t **test:** The test used with a two-class, categorical independent variable and a continuous dependent variable.

Theoretical prediction: A prediction, grounded in logic, about whether or not a certain event will occur.

Theory: A set of proposed and testable explanations about reality that are bound together by logic and evidence.

Trends: Patterns that indicate whether something is increasing, decreasing, or staying the same over time.

Trial: An act that has several different possible outcomes.

Tukey's honest significant difference: A widely used post hoc test used in ANOVA when the null is rejected as a means of determining the number and location of differences between groups.

Two-tailed tests: Hypothesis tests in which alpha is split in half and placed in both tails of the distribution such that there are two values of the test statistic. Also called *nondirectional tests.*

Type I error: The erroneous rejection of a true null hypothesis.

Type II error: The erroneous retention of a false null hypothesis.

Unit of analysis: The object or target of a research study.

Univariate: Involving one variable.

Variable: A characteristic that takes on multiple values in a sample or population.

Variation ratio: A measure of dispersion for variables of any level of measurement that is calculated as the proportion of cases located outside the modal category.

Within-group variance: The amount of diversity that exists among the people or objects in a single group or class. This is a measure of random fluctuation, or error.

z **score:** A standardized version of a raw score that offers two pieces of information about the raw score: (1) how close it is to the distribution mean, and (2) whether it is greater than or less than the mean.

z **table:** A table containing a list of z scores and the area of the curve that is between the distribution mean and each individual z score.

ANSWERS TO REVIEW QUESTIONS AND PROBLEMS

Chapter 1

1. Science is a systematic and controlled way of gathering information about the world. Methods are integral to science because scientific results are only trustworthy when the procedures used to reach them are considered correct by others in the scientific community.

3. A population is the universe of people, places, or objects that a researcher wishes to study. Populations are usually too large to be studied directly.

5. Replication is important in science because the emphasis on methods means that one person's findings should be able to be replicated by another person using similar methods. It is also important to examine findings using different methods to see if the results remain stable.

Chapter 2

1.
   a) education
   b) crime
   c) individuals

3.
   a) poverty
   b) violent crime
   c) neighborhoods

5.
   a) money spent on education, health, and welfare
   b) violent crime
   c) countries

7.
   a) police department location
   b) starting pay for officers
   c) police departments

9. The omitted variable bias occurs when an independent variable that is strongly related to a dependent variable has been left out of a statistical analysis. This can lead to erroneous conclusions about the factors that influence the dependent variable that is being studied.

11.

  a) nominal
  b) interval
  c) nominal
  d) ratio
  e) ratio
  f) ordinal
  g) nominal

13.

  a) ratio, ordinal, nominal
  b) Possible phrasing 1 is the best because it will create a ratio-level variable, and it is always best to use the highest level of measurement that is possible in any given situation. There are more analytic techniques available for ratio-level variables than for nominal- and ordinal-level ones.

15.

  a) nominal
  b) ratio

17.

  a) the presence of a victim advocacy office in the courthouse
  b) nominal
  c) offender sentencing
  d) ratio

19.

  a) city homicide rates
  b) ratio
  c) firearm ownership
  d) nominal

Chapter 3

1. a.

p	pct	cf	cp	c%
.07	6.61	745	.07	6.61
.06	6.09	1,431	.13	12.70
.87	87.30	11,268	1.00	100.00

  b. The appropriate display method would be a pie chart or a bar graph.

c.

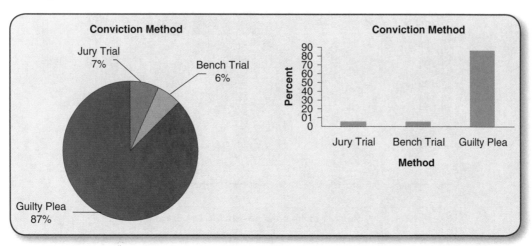

3. a. Histogram for ungrouped data

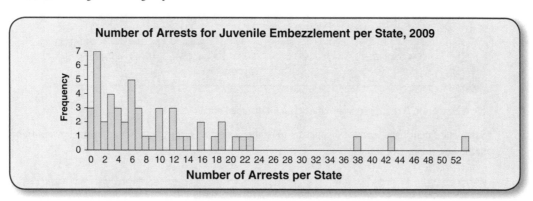

b. The range is $53 - 0 = 53$. With 10 intervals, the width of each would be $width = \dfrac{53}{10} = 5.30$. Rounding to the nearest whole number, the width is 5.00. (Note: The grouping process results in 11, rather than 10, intervals because the intervals must contain all of the original values. It is therefore necessary to use an additional interval to capture the `53' at the end of the data set.)

Stated Class Limits	f
0 – 4	19
5 – 9	12
10 – 14	8
15 – 19	5
20 – 24	3

(Continued)

(Continued)

Stated Class Limits	f
25 – 29	0
30 – 34	0
35 – 39	1
40 – 44	1
45 – 49	0
50 – 54	1
	$N = 50$

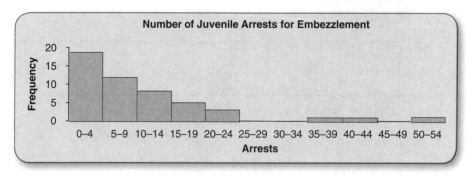

c. A histogram is the appropriate chart type for grouped data.

5. Support for marijuana legalization appears to be increasing over time, as indicated by the upward trend in the line graph.

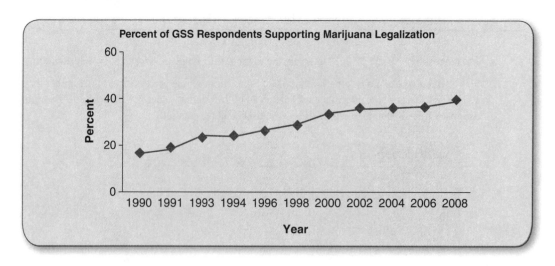

7. a.

City	Violent Crime Rate
Sacramento	8.86
Beverly Hills	2.35
Paso Robles	3.18
Bakersfield	6.34
San Francisco	7.36
Los Angeles	6.25
San Bernardino	9.56

b.

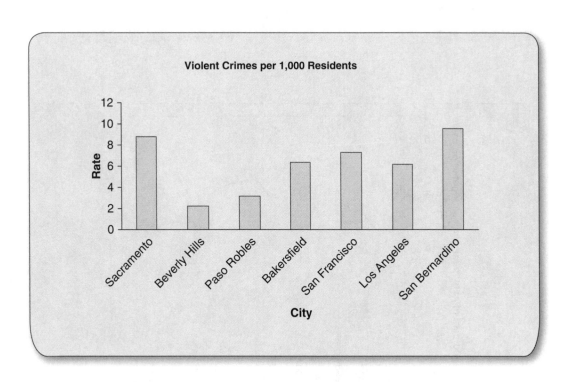

9.

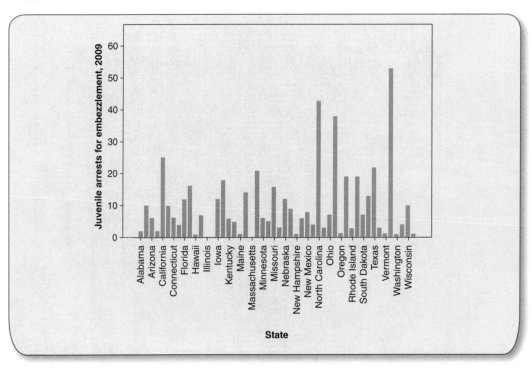

11.

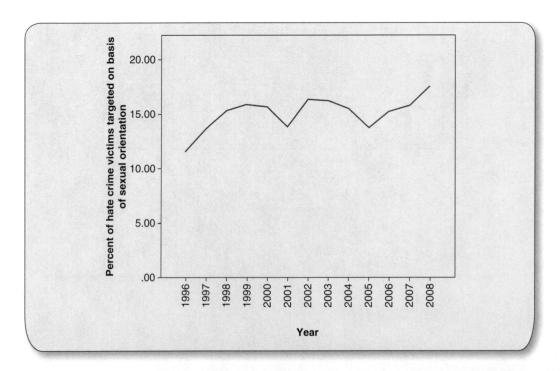

Chapter 4

1.
   a) ratio
   b) mode, median, mean

3.
   a) nominal
   b) mode

5. The mean is the midpoint of the magnitudes because it is the arithmetic "fulcrum" of a set of scores. The mean perfectly balances positive and negative deviation scores. Deviation scores, computed as $d_i = x_i - \overline{x}$, measure each raw score's distance from the distribution mean. Deviation scores sum to zero.

7. c

9. a

11.
   a) 0
   b) acquaintance

13.
   a) 0
   b) .11

15.
   a) 239
   b) 265.00
   c) The mean is greater than the median, so this distribution is positively skewed.

17.
   a) mean = 6.2051, and median = 5.6000
   b) The mean is greater than the median, so this distribution is positively skewed.

Chapter 5

1. .46

3. .42

5. .30

7. 66.75

9. 2.42

11. .26

13. .17

15. 189.60

17. 167.12

19. The standard deviation is the mean of the deviation scores.

21.

Statistics

Percent of sworn full-time officers that is female

N	Valid	71
	Missing	0
Mean		10.6638
Median		10.8280
Mode		.00
Std. Deviation		4.58288
Variance		21.003
Range		22.12

Chapter 6

1.
   a) 8
   b) 2
   c) 28

3.
   a) 7
   b) 3
   c) 35

5.
   a) 8
   b) 4
   c) 70

7. standard normal distribution

9. standard normal distribution

11. binomial distribution

13.

r	$\binom{N}{r}$	p^r	q^{N-r}	$p(r)$
0	$\binom{6}{0}=1$	$.70^0 = 1.00$	$.30^{6-0=6}=.001$	.001
1	$\binom{6}{1}=6$	$.70^1 = .70$	$.30^{6-1=5}=.002$	.01
2	$\binom{6}{2}=15$	$.70^2 = .49$	$.30^{6-2=4}=.01$	.07
3	$\binom{6}{3}=20$	$.70^3 = .34$	$.30^{6-3=3}=.03$	.20
4	$\binom{6}{4}=15$	$.70^4 = .24$	$.30^{6-4=2}=.09$	.32
5	$\binom{6}{5}=6$	$.70^5 = .17$	$.30^{6-5=1}=.30$	.31
6	$\binom{6}{6}=1$	$.70^6 = .12$	$.30^{6-6=0}=1.00$	.12

b) $r=4$

c) $r=0$

d) .08

e) .43

15.

r	$\binom{N}{r}$	p^r	q^{N-r}	$p(r)$
0	$\binom{5}{0}=1$	$.59^0 = 1.00$	$.41^{5-0=5}=.01$	.01
1	$\binom{5}{1}=5$	$.59^1 = .59$	$.41^{5-1=4}=.03$	.09
2	$\binom{5}{2}=10$	$.59^2 = .35$	$.41^{5-2=3}=.07$	.25
3	$\binom{5}{3}=10$	$.59^3 = .21$	$.41^{5-3=2}=.17$	.36
4	$\binom{5}{4}=5$	$.59^4 = .12$	$.41^{5-4=1}=.41$	.25
5	$\binom{5}{5}=1$	$.59^5 = .07$	$.41^{5-5=0}=1.00$	.07

    b) $r=3$
    c) $r=0$
    d) .10
    e) .32

17.
    a) 3.08
    b) .4990
    c) .001

19.
    a) .24
    b) .0948
    c) .4052

21. Values of $z \geq 1.89$

23. Values of $z \geq 1.29$

25. Values of $z \leq -1.29$

Chapter 7

1. a

3. b

5. sampling

7. b

9. z

11. a

13. c

15. b

Chapter 8

1. 100

3. z

5. The z distribution cannot be used with small samples because the central limit theorem does not guarantee normality of the sampling distribution when sample size is low; therefore, the t distribution must be used because it can accommodate small samples.

7. 95% *CI*: $1.56 \leq \mu \leq 1.64$

   There is a 95% chance that the interval 1.56 to 1.64, inclusive, contains the population mean.

9. 99% *CI*: $.58 \leq P \leq .78$

   There is a 99% chance that the interval .58 to .78, inclusive, contains the population proportion.

11. 95% *CI*: $.58 \leq P \leq .62$

    There is a 95% chance that the interval .58 to .62, inclusive, contains the population proportion.

13. 99% *CI*: $.43 \leq P \leq .63$

    There is a 99% chance that the interval .43 to .53, inclusive, contains the population proportion.

15. 99% *CI*: $0.41 \leq \mu \leq 0.99$

    There is a 99% chance that the interval 0.41 to 0.99, inclusive, contains the population mean.

17. 95% *CI*: $.51 \leq P \leq .55$

There is a 95% chance that the interval .51 to .55, inclusive, contains the population proportion.

19. 99% *CI*: $43.56 \leq \mu \leq 48.72$

There is a 99% chance that the interval 43.56 to 48.72, inclusive, contains the population mean.

21. 99% *CI*: $45.79 \leq \mu \leq 64.21$

There is a 99% chance that the interval 45.79 to 64.21, inclusive, contains the population mean.

Chapter 9

1. One possible explanation is that the difference is a product of sampling error and is therefore meaning-less. The other potential reason for the difference is that there truly is a "gender effect" and male and female defendants receive disparate sentences.

3. The alternative (symbolized H_1) predicts that there is a relationship between the independent and dependent variables; in other words, any observed or empirical indications that there might be a relationship indicate a true underlying association.

5. A Type I error occurs when a true null is wrongly rejected; that is, when there actually is no relationship between two variables in the population, but the results of a hypothesis test in a sample lead to the erroneous conclusion that the variables are related.

7. The word *bivariate* refers to the presence of two variables. Bivariate hypothesis tests, for instance, are tests involving one independent variable and one dependent variable.

9. A low probability would lead to the conclusion that this empirical outcome is an atypical, unexpected finding. The implication is that the empirical result is the product of a true effect, a genuine relationship between the two variables under examination.

11. Chi-square

13. Chi-square

Chapter 10

1.

a) IV: Whether a state's violent crime rate is high or low, and DV: Whether a state authorizes the death penalty or not
b) Both variables are nominal
c)
<u>Step 1:</u> $H_0: \chi^2 = 0, H_1: \chi^2 > 0$
<u>Step 2:</u> χ^2 distribution with $df = 1$

<u>Step 3:</u> $\chi^2_{crit} = 6.635$ and decision rule is *If $\chi^2_{obt} > 6.635$, H_0 will be rejected*
<u>Step 4:</u> $\chi^2_{obt} = 4.27$
<u>Step 5:</u> Decision: χ^2_{obt} is not greater than 6.635, so the null is retained. Conclusion: A state's violent crime rate does not affect whether or not that state authorizes the death penalty.

3.

    a) IV: Respondents' driving frequency, and DV: Whether or not respondents had experienced a traffic stop in the past 12 months
    b) The IV is ordinal and the DV is nominal
    c)

<u>Step 1:</u> H_0: $\chi^2 = 0$, H_1: $\chi^2 > 0$
<u>Step 2:</u> χ^2 distribution with $df = 2$
<u>Step 3:</u> $\chi^2_{crit} = 9.210$ and decision rule is *If $\chi^2_{obt} > 9.210$, H_0 will be rejected*
<u>Step 4:</u> $\chi^2_{obt} = 3.67$
<u>Step 5:</u> Decision: χ^2_{obt} is not greater than 9.210, so the null is retained. Conclusion: Among black males, driving frequency does not affect the likelihood of having experienced a recent traffic stop.

5.

    a) IV: The number of stops a person has experienced recently, and DV: The number of times someone has called the police to report problems
    b) Both variables are ordinal
    c)

<u>Step 1:</u> H_0: $\chi^2 = 0$, H_1: $\chi^2 > 0$
<u>Step 2:</u> χ^2 distribution with $df = 2$
<u>Step 3:</u> $\chi^2_{crit} = 13.815$ and decision rule is *If $\chi^2_{obt} > 13.815$, H_0 will be rejected*
<u>Step 4:</u> $\chi^2_{obt} = 339.08$
<u>Step 5:</u> Decision: χ^2_{obt} is greater than 13.815, so the null is rejected. Conclusion: People's recent experiences with traffic stops do affect the likelihood that they would call the police to report problems.

7.

    a) IV: Prison type, and DV: Whether or not a prison offers college courses
    b) Both variables are nominal
    c)

<u>Step 1:</u> H_0: $\chi^2 = 0$, H_1: $\chi^2 > 0$
<u>Step 2:</u> χ^2 distribution with $df = 1$
<u>Step 3:</u> $\chi^2_{crit} = 6.635$ and decision rule is *If $\chi^2_{obt} > 6.635$, H_0 will be rejected*
<u>Step 4:</u> $\chi^2_{obt} = 62.13$
<u>Step 5:</u> Decision: χ^2_{obt} is greater than 6.635, so the null is rejected. Conclusion: Private and public prisons do differ with respect to the offering of college courses to inmates.

9.

    a) IV: Defendant gender, and DV: Conviction offense
    b) Both variables are nominal
    c)

<u>Step 1:</u> H_0: $\chi^2 = 0$, H_1: $\chi^2 > 0$
<u>Step 2:</u> χ^2 distribution with $df = 3$

<u>Step 3:</u> $\chi^2_{crit} = 7.815$ and decision rule is *If* $\chi^2_{obt} > 7.815$, H_0 *will be rejected*
<u>Step 4:</u> $\chi^2_{obt} = 120.39$
<u>Step 5:</u> Decision: χ^2_{obt} is greater than 7.815, so the null is rejected. Conclusion: Defendants' gender is related to the type of crime for which they are convicted.

11.

    a) IV: Respondent gender, and DV: Perceptions of stop legitimacy
    b) Both variables are nominal
    c)

<u>Step 1:</u> $H_0: \chi^2 = 0, H_1: \chi^2 > 0$
<u>Step 2:</u> χ^2 distribution with $df = 1$
<u>Step 3:</u> $\chi^2_{crit} = 3.841$ and decision rule is *If* $\chi^2_{obt} > 3.841$, H_0 *will be rejected*
<u>Step 4:</u> $\chi^2_{obt} = 2.34$
<u>Step 5:</u> Decision: χ^2_{obt} is not greater than 3.841, so the null is retained. Conclusion: There is no relationship between drivers' gender and their perceptions of the legitimacy of traffic stops.

13.

    a) $\chi^2_{obt} = 4.908$
    b) The significance level is .027, which is less than .05, so the null is rejected
    c) Defendants' gender is related to their likelihood of pleading guilty versus going to trial
    d) Lambda is 0.00 and Cramer's V is .023, so this is a very weak relationship.

15.

    a) $\chi^2_{obt} = 43.775$
    b) the significance level is .000, which is less than .01, so the null is rejected
    c) There is a relationship between prison security level and the escape rate
    d) Somers' d is $-.098$, which is statistically significant ($p = .000$) but very small; gamma is $-.671$ ($p = .000$), which indicates a fairly robust relationship. They are both negative, indicating that higher security levels are associated with fewer escapes. While there seems to be an association, then, security level does not help predict facilities' escape rates.

Chapter 11

1.

    a) pleading guilty versus going to trial
    b) nominal
    c) months of incarceration
    d) ratio (this sample includes defendants who were not sentenced to incarceration)

3. b

5.

<u>Step 1:</u> $H_0: \mu_1 = \mu_2$, $H_1: \mu_1 \neq \mu_2$
<u>Step 2:</u> t distribution with $df = 616$

Note: Answers to review problems in this and subsequent chapters may vary depending on the number of steps used and whether or not rounding is employed during the calculations. The answers presented here were derived using the procedures illustrated in the main text.

<u>Step 3:</u> $t_{crit} = \pm 1.980$ (± 1.960 would likely also be acceptable) and decision rule is *If t_{obt} is either > 1.980 or < −1.980, H_0 will be rejected*
<u>Step 4:</u> $\hat{\sigma}_{\hat{p}_1 - \hat{p}_2} = .03$ and $t_{obt} = -.33$
<u>Step 5:</u> Decision: t_{obt} is not less than −1.980, so the null is retained. Conclusion: Multiple homicide offenders do not differ from single homicide offenders in terms of diversity of offending.

7.

<u>Step 1:</u> $H_0: \mu_1 = \mu_2, H_1: \mu_1 > \mu_2$
<u>Step 2:</u> t distribution with $df = 159$
<u>Step 3:</u> $t_{crit} = 2.358$ and decision rule is *If t_{obt} is > 2.358, H_0 will be rejected*
<u>Step 4:</u> $\hat{\sigma}_{\hat{p}_1 - \hat{p}_2} = .58$ and $t_{obt} = 4.91$
<u>Step 5:</u> Decision: $t_{obt} > 2.358$, so the null is rejected. Conclusion: Police agencies in urban jurisdictions have significantly higher percentages of female officers relative to agencies in nonurban areas.

9.

<u>Step 1:</u> $H_0: \mu_1 = \mu_2, H_1: \mu_1 \neq \mu_2$
<u>Step 2:</u> t distribution with $df = 131.16 \approx 131$
<u>Step 3:</u> $t_{crit} = \pm 2.617$ and decision rule is *If t_{obt} is either > 2.617 or < −2.617, H_0 will be rejected*
<u>Step 4:</u> $\hat{\sigma}_{\hat{p}_1 - \hat{p}_2} = 14.73$ and $t_{obt} = -1.79$
<u>Step 5:</u> Decision: t_{obt} is neither > 2.617 nor < −2.617, so the null is retained. Conclusion: The mean jail sentence for juveniles who were under 16 at the age of arrest does not differ significantly from that for juveniles who were over age 16.

11.

<u>Step 1:</u> $H_0: \mu_1 = \mu_2, H_1: \mu_1 < \mu_2$
<u>Step 2:</u> t distribution with $df = 5$
<u>Step 3:</u> $t_{crit} = -3.365$ and decision rule is If t_{obt} is < −3.365, H_0 will be rejected
<u>Step 4:</u> $s_D = 3.74$ and $\sigma_{\hat{p}_1 - \hat{p}_2} = 1.53$ and $t_{obt} = -1.07$
<u>Step 5:</u> Decision: t_{obt} is not < −3.365, so the null is retained. Conclusion: Homicides in large cities in New York did not significantly increase after the abolition of the death penalty.

13.

<u>Step 1:</u> $H_0: P_1 = P_2, H_1: P_1 \neq P_2$
<u>Step 2:</u> z distribution
<u>Step 3:</u> $z_{crit} = \pm 1.96$ and decision rule is *If z_{obt} is either > 1.96 or < −1.96, H_0 will be rejected*
<u>Step 4:</u> $\hat{\sigma}_{\hat{p}_1 - \hat{p}_2} = .07$ and $z_{obt} = 4.57$
<u>Step 5:</u> Decision: z_{obt} is > 1.96, so the null is rejected. Conclusion: The proportion of released juvenile drug defendants represented by public attorneys whose cases are disposed of within 180 days is significantly different from those juvenile defendants represented by private attorneys.

15. This is dependent samples t.

   a) $t_{obt} = -2.226$, which is very similar to what we obtained by hand
   b) The null would be rejected at an alpha level of .05 because the significance level is .026 and .026 < .05.
   c) There is a statistically significant difference between daytime and nighttime stops in terms of the number of officers present. According to the "Paired Sample Statistics" table, the mean number of officers present during daytime stops is 1.66, which is slightly higher than the mean of 1.45 present during nighttime stops.

Chapter 12

1.
   a) judges' gender
   b) nominal
   c) sentencing decisions
   d) ratio
   e) two-population t test

3.
   a) arrest
   b) nominal
   c) domestic violence recidivism
   d) ratio
   e) ANOVA

5.
   a) poverty
   b) ordinal
   c) crime
   d) ratio
   e) ANOVA

7.
   Step 1: H_0: $\mu_1 = \mu_2 = \mu_3$, H_1: some $\mu_i \neq$ some μ_j
   Step 2: F distribution with $df_B = 2$ and $df_W = 18$
   Step 3: $F_{crit} = 3.55$ and decision rule is If $F_{obt} > 3.55$, H_0 will be rejected
   Step 4: $F_{obt} = 3.78$
   Step 5: Decision: $F_{obt} > 3.55$, so the null is rejected. Conclusion: There is a significant difference across crime types in terms of the number of wiretaps authorized in 2009; that is, there is a relationship between crime type and the number of wiretaps authorized. $\omega^2 = .21$, meaning crime type accounts for 21% of the variance in wiretaps.

9.
   Step 1: H_0: $\mu_1 = \mu_2 = \mu_3 = \mu_4$, H_1: some $\mu_i \neq$ some μ_j
   Step 2: F distribution with $df_B = 3$ and $df_W = 45$
   Step 3: $F_{crit} = 4.31$ and decision rule is If $F_{obt} > 4.31$, H_0 will be rejected
   Step 4: $F_{obt} = 1.91$
   Step 5: Decision: F_{obt} is not greater than 4.31, so the null is retained. Conclusion: There is no significant difference across region in the percentage of assaults against police officers that are committed with firearms.

11.
   Step 1: H_0: $\mu_1 = \mu_2 = \mu_3$, H_1: some $\mu_i \neq$ some μ_j
   Step 2: F distribution with $df_B = 2$ and $df_W = 27$
   Step 3: $F_{crit} = 3.35$ and decision rule is If $F_{obt} > 3.35$, H_0 will be rejected
   Step 4: $F_{obt} = .35$
   Step 5: Decision: F_{obt} is not greater than 3.42, so the null is retained. Conclusion: The gender of inmates held in a prison does not affect the inmate-on-inmate assault rate in that institution.

13.

  b) $F_{obt} = 13.931$

  c) The null would be rejected at $\alpha = .01$ because $p = .000$

  d) Attorney type is related to the number of days adult drug defendants are held in jail before being released pending trial

  e) Tukey's and the Bonferroni both show that assigned counselors stand out as having the highest mean number of days-to-release for their clients; assigned counselors differ significantly from both public and private attorneys, while the difference between public defenders and private attorneys is not significant.

  f) $\omega^2 = .01$, so just 1% of the variance in days to pretrial release for drug defendants is attributable to attorney type.

Chapter 13

1.

  a) parental incarceration

  b) nominal

  c) adult children's incarceration

  d) nominal

  e) chi-square

3.

  a) socioeconomic disadvantage

  b) ratio

  c) violent crime

  d) ratio

  e) correlation

5. A linear relationship is one in which the amount and direction of change in the dependent variable that is associated with a one-unit increase in the independent variable remains constant across all levels of the independent variable. This type of relationship can be represented by a straight line with a constant slope.

7.

  Step 1: $H_0: \rho = 0, H_1: \rho \neq 0$

  Step 2: t distribution with $df = 3$

  Step 3: $t_{crit} = \pm 3.182$ and decision rule is *If t_{obt} is either > 3.182 or < −3.182, H_0 will be rejected*

  Step 4: $r = .77$ and $t_{obt} = 2.09$

  Step 5: Decision: t_{obt} is neither > 3.182 nor < −3.182, so the null is retained. Conclusion: There is no correlation between the number of dollars states spend on prisons per capita and the violent crime rates in those states.

9.

  Step 1: $H_0: \rho = 0, H_1: \rho > 0$

  Step 2: t distribution with $df = 7$

  Step 3: $t_{crit} = 1.895$ and decision rule is *If t_{obt} is > 1.895, H_0 will be rejected*

<u>Step 4:</u> $r = .75$ and $t_{obt} = 2.99$

<u>Step 5:</u> Decision: t_{obt} is > 1.895, so the null is rejected. Conclusion: There is a positive correlation between the number of crimes in an area and the number of police agencies in that location. Sign: positive; Magnitude: strong; Coefficient of determination = .56, meaning 56% of the variance in number of police agencies per 1,000 square miles is due to crimes per square mile.

11.

  <u>Step 1:</u> H_0: $\rho = 0$, H_1: $\rho < 0$

  <u>Step 2:</u> t distribution with $df = 8$

  <u>Step 3:</u> $t_{crit} = -2.896$ and decision rule is *If t_{obt} is < -2.896, H_0 will be rejected*

  <u>Step 4:</u> $r = .74$ and $t_{obt} = 3.12$

  <u>Step 5:</u> Decision: t_{obt} is not less than -2.896, so the null is retained. Conclusion: Handgun and knife murder rates are not negatively correlated.

13.

  b) $r = .082$

  c) The null would not be rejected at $\alpha = .05$ because $p = .628$

  d) There is no correlation between the number of people a state executes in one year and that state's homicide rate the following year

  e) The null was not rejected, so interpretation is not appropriate.

Chapter 14

1.

  a) $b = -.05$

  b) $a = 4.04$

  c) $\hat{Y} = 4.04 - .05x$

  d) $\hat{Y}_3 = 3.89$, $\hat{Y}_{15} = 3.29$

  e)

  <u>Step 1:</u> H_0: $B = 0$, H_1: $B \neq 0$

  <u>Step 2:</u> t distribution with $df = 3$

  <u>Step 3:</u> $t_{crit} = \pm 3.182$, decision rule is that *If t_{obt} is either > 3.182 or < -3.182, H_0 will be rejected*

  <u>Step 4:</u> $SE_b = .27$, $t_{obt} = -.19$

  <u>Step 5:</u> Decision: $-.19$ is neither > 3.182 nor < -3.182, so the null is retained. Conclusion: The number of prior incarcerations that inmates have experienced is not a significant predictor of their in-prison behavior

  f) N/A, as the null was retained.

3.

  a) $b = 1.08$

  b) $a = -3.57$

  c) $\hat{Y} = -3.57 + 1.08x$

  d) $\hat{Y}_4 = .75$, $\hat{Y}_8 = 5.07$

  e)

  <u>Step 1:</u> H_0: $B = 0$, H_1: $B \neq 0$

  <u>Step 2:</u> t distribution with $df = 8$

  <u>Step 3:</u> $t_{crit} = \pm 2.306$, decision rule is that *If t_{obt} is either > 2.306 or < -2.306, H_0 will be rejected*

<u>Step 4:</u> $SE_b = .30$ and $t_{obt} = 3.60$

<u>Step 5:</u> Decision: $3.60 > 2.306$, so the null is rejected. Conclusion: There is a statistically significant relationship between unemployment and violent crime

f) $\beta = .79$.

5.

    a) $b = .63$

    b) $a = -1.76$

    c) $\hat{Y} = -1.76 + .63x$

    d) $\hat{Y}_5 = 1.39$, $\hat{Y}_{10} = 4.54$

    e)

<u>Step 1:</u> $H_0: B = 0, H_1: B \neq 0$

<u>Step 2:</u> t distribution with $df = 8$

<u>Step 3:</u> $t_{crit} = \pm 2.306$, decision rule is that *If t_{obt} is either > 2.306 or < −2.306, H_0 will be rejected*

<u>Step 4:</u> $SE_b = .18$ and $t_{obt} = 3.50$

<u>Step 5:</u> Decision: $3.50 > 2.306$, so the null is rejected. Conclusion: There is a statistically significant relationship between poverty and violent crime

f) $\beta = .77$.

7. [SPSS output]

9. $\hat{Y} = 115.17 - 4.69x$

11.

    a) The null hypothesis for *age* would be rejected because its *p* value is .000, which is less than .05.

    b) The null hypothesis for *charges* would be rejected because its *p* value is .000, which is less than .05.

13. $\beta_{age} = -.126$ and $\beta_{charge} = .103$; these are substantively weak in magnitude.

15. [SPSS output]

17. $\hat{Y} = -.41 + .64x$

19. The null is not rejected for SNAP because its *p* value is .078, which is greater than .05; the null is rejected for unemployment because $.003 < .05$; the null is rejected for poverty because $.031 < .05$.

21. $\beta_{unemployement} = .56$ and $\beta_{poverty} = .45$; these are both fairly strong in magnitude, with the beta for unemployment being quite robust.

INDEX

⑤SAGE research methods online

The essential tool for researchers . . .

. . . from the world's leading research methods publisher

Discover SRMO Lists—methods readings suggested by other SRMO users

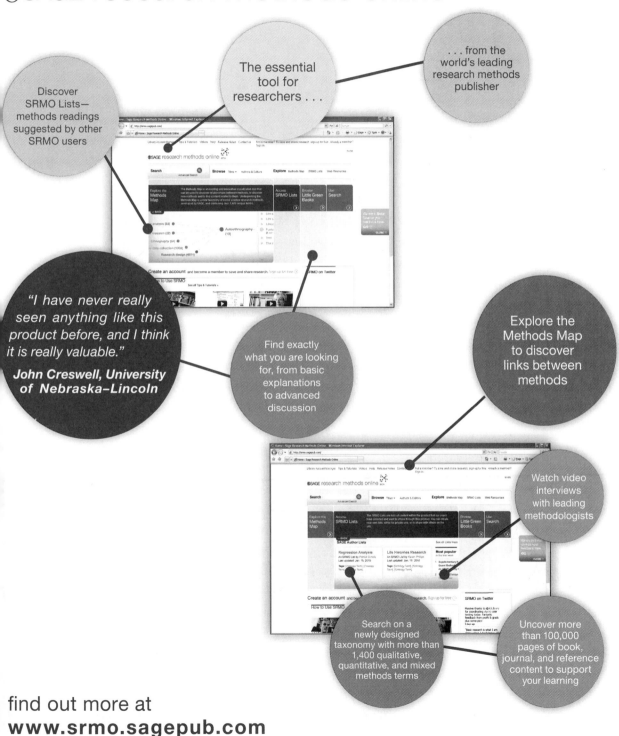

"I have never really seen anything like this product before, and I think it is really valuable."

John Creswell, University of Nebraska–Lincoln

Find exactly what you are looking for, from basic explanations to advanced discussion

Explore the Methods Map to discover links between methods

Watch video interviews with leading methodologists

Search on a newly designed taxonomy with more than 1,400 qualitative, quantitative, and mixed methods terms

Uncover more than 100,000 pages of book, journal, and reference content to support your learning

find out more at
www.srmo.sagepub.com